WHOOSΠ →BOOM← Splat

BUILD JAM JAR JETS, ELASTIC ZIP CANNONS, CLOTHESPIN SNAP SHOOTERS, *AND MORE* LEGENDARY LAUNCHERS

REVISED AND EXPANDED SECOND EDITION

WILLIAM GURSTELLE

CHICAGO REVIEW PRESS

Copyright © 2017 by William Gurstelle
All rights reserved
Revised and updated second edition
Published by Chicago Review Press Incorporated
814 N. Franklin Street
Chicago, Illinois 60610
978-1-61373-907-5

First edition copyright © 2007 by William Gurstelle was published by Three Rivers Press, an imprint of the Crown Publishing Group, a division of Random House, Inc., New York.

Library of Congress Cataloging-in-Publication Data
Names: Gurstelle, William, author.
Title: Whoosh boom splat : build jam jar jets, elastic zip cannons,
    clothespin snap shooters, and more legendary launchers / William Gurstelle.
Description: Revised and updated second edition. | Chicago, Illinois :
    Chicago Review Press Incorporated, [2017] | Includes bibliographical
    references and index.
Identifiers: LCCN 2017012444 (print) | LCCN 2017018724 (ebook) | ISBN
    9781613739082 (pdf) | ISBN 9781613739105 (epub) | ISBN 9781613739099
    (mobipocket) | ISBN 9781613739075 | ISBN 9781613739075(pbk.) | ISBN
    1613739079(pbk.)
Subjects: LCSH: Science—Experiments. | Ballistics—Experiments. | Toy
    making. | Handicraft. | Miniature weapons.
Classification: LCC TT174 (ebook) | LCC TT174 .G87 2017 (print) | DDC
    745.592—dc23
LC record available at https://lccn.loc.gov/2017012444

Cover design: Andrew Brozyna
Interior design: Jonathan Hahn
Interior illustrations: Damien Scogin

Printed in the United States of America
5 4 3 2 1

To Karen, my helper, consultant, and inspiration.
She makes my heart go whoosh *and* boom.

CONTENTS

Preface to the Second Edition

We've seen so, so many changes since *Whoosh Boom Splat* was first published in 2006. There were no smartphones. People mostly bought music in physical formats like compact discs. There were Radio Shack stores in nearly every shopping mall. But one thing hasn't changed—people still enjoy making things that go whoosh or splat. That will never change, because there's something innately interesting and satisfying about designing and making things yourself, and, well, a lot of people just like the idea of making things go boom.

So, I've updated this book. You will find several new projects in it that I hope you'll find interesting. In addition, the original projects have been updated and the graphics have been redone to make them fresher, more clear, and more interesting.

INTRODUCTION

Since my first book, *Backyard Ballistics*, was published, a couple things have happened. First, the number of books in my library devoted to researching things that go *whoosh, boom,* and *splat* has grown enormously. Concomitantly, the bookshelf holding biographies of inventors has also grown quickly.

Many of the scientists and practical philosophers associated with this field of study are pretty well known. For example, books about Alfred Nobel (the inventor of dynamite), Robert Goddard (the father of liquid fuel rocketry), and Archimedes (the greatest of the ancient Greek scientists and discoverer of the principle of buoyancy) are easy to find. But many more scientific innovators are not as well known as they should be. In this book you will meet several scientists and historical figures who perhaps should be more famous, considering how adept they were at making things go boom and how the events that followed their work changed the course of history.

The names Urban of Hungary, Federico Gianibelli, and Waldo Semon are just about forgotten, and even more important worldchanging scientists like Alexander von Humboldt and Count Rumford are familiar to only a few. These men were able experimenters and terrific scientists who shaped wood, cut metal, mixed chemicals, and occasionally got into trouble. But they also did great things.

The other thing that's happened since the publication of *Backyard Ballistics* is that people often ask me, "Are you the fellow who wrote that book about blowing things up?"

Well, golly—no, I'm not.

I never really think of *Backyard Ballistics* as being about blowing things up. It is not a cookbook for anarchists, and there are no recipes in it for making bombs. The projects in *Backyard Ballistics* do result in things going boom, at least occasionally. But is that a good thing? Is there anything inherently good and noble about experimentation and fabrication, when the desired result is nothing more than a really good bang? I myself wasn't sure until I recalled something a friend told me several years ago.

This particular friend, Jim, is a tall, laconic electrical engineer, a product of the northern plains. He grew up in a rural area in northwestern Minnesota, around the towns of Red Lake, Fertile, and Climax. As a boy, he spent his free time taking apart farm machinery and gadgets and then putting them back together. When he was old enough, he joined the air force, hoping to see the world. In the irony of military life, Jim actually wound up seeing more of the northern prairie—he was assigned to a program targeting instructions for the nuclear missiles that were then hidden underneath North Dakota.

During the 1960s Jim tinkered on the hundreds of Minuteman intercontinental ballistic missiles buried in pop-top tombs under the winter wheat fields. At the time of his enlistment there were so many rockets up there that if North Dakota had seceded from the Union, it would have become the world's second-largest nuclear power.

Jim's work involved targeting nuclear missiles by straddling the nose cone at the tip of the rocket and shooting the coordinates of visible landmarks through a transit-like device called a theodolite. Then he fed a metal punch tape into the tape-reader fronting the missile guidance system in the communications bay of the rocket. The tape provided the exact coordinates of that week's top Cold War target—say, the Soviet Strategic Rocket Force bases at Tatischevo, or the Northern Fleet Command bunker at Severomorsk. This told the sleeping Minuteman exactly where it should head, should the go-code arrive from the Pentagon to wake it up.

From this background of agrarian and military mechanics, Jim took the logical step of enrolling in college after his discharge. Because he enjoys figuring out how things work, he earned his degree in engineering.

The love of tinkering and a concomitant affection for making things go whoosh, or zoom, or even boom, should be the identification badge by which modern engineers are recognized. "The greater a person's tinkering skills," Jim told me, "the greater the engineer."

Jim observed that his fellow rural and small-town boys (they were mostly boys back then) in the mid-20th century who filled their free time with tinkering got good at it. Many who left the farmlands followed the very same path to study engineering as my friend Jim did. They had a practical basis for their knowledge and study, rooted in long hours of working on things with their hands and, occasionally, blowing something up—not violently, not dangerously, but on purpose and with care. They became good, even great, engineers.

These engineers looked at themselves not just as technicians or professional designers but as engineer-artists whose creations were both mechanical art and extensions of themselves.

To predict the adult engineering success of a youth, look to his or her propensity for making things go *whoosh, boom,* or *splat.* Thomas Edison's childhood chemistry experiments, performed in the baggage car of the train on which he worked, resulted in the accidental destruction of the railroad car. B. F. Skinner, the great social scientist, was proud of the steam cannons and catapults he built as a youngster. Homer Hickam, the NASA engineer who wrote the bestselling book *Rocket Boys,* predictably enough burned down part of his yard with primitive rocket experiments. A certain Thomas Shelton, of whom we'll learn more, accidentally demolished his parents' kitchen with an experiment gone wrong. And a young engineer named Jack Welch burned down one of his employer's factories, yet managed to retain his job. He went on to become the CEO of General Electric.

Making things go *whoosh, boom,* and *splat* is a rite of passage for the intellectually curious. Most of the time such events don't happen on purpose, but happen they do. This book does not advocate reckless experimentation. Danger and recklessness are the antithesis of good experimental technique. I still remember one particularly pithy comment Jim made to me many years ago: "Science isn't just about blowing things up," he told me. "Rather, it's about blowing things up and knowing how you did it."

So with this in mind, read the next chapter carefully.

FIRST: SAFETY

When you were a child, people told you not to play with matches for a good reason—they can be dangerous! If you don't follow the directions closely, any of the projects in *Whoosh Boom Splat* could cause harm to you and your possessions. Remember to always follow the instructions closely. Do not make changes to the materials or construction techniques. It can lead to unexpected and unintended results.

> ## A Very Important Message
>
> The projects described in the following pages have been designed with safety foremost in mind. However, as you try them, something unexpected may still possibly occur. It is important that you understand that neither the author, the publisher, nor the bookseller can or will guarantee your safety. When you try the projects described here, *you do so at your own risk*.

Some of the projects included in this book have been popular for many years while others are new. Unfortunately, in rare instances, damage to both property and person occurred when something went wrong. The likelihood of such an occurrence is remote, as long as the directions are followed, but remember this: *things can go wrong*. Always use common sense and remember that all experiments and projects are carried out at your own risk.

Be aware that each city, town, or municipality has its own rules and regulations, some of which may apply to the projects described in *Whoosh Boom Splat*. Further, local authorities have wide latitude to interpret the law. Therefore, you should take the time to learn the rules, regulations, and laws of the area in which you plan to carry out these projects. A check with local law enforcement will tell you whether the project is suitable for your area. If not, there are plenty of other places where experiments can be undertaken safely and legally. If in doubt, be sure to check first!

GROUND RULES

These are your general safety rules. Each project will also provide specific safety instructions.

1. The projects described here run the gamut from simple to complex. All are designed to be done by adults or, at a minimum, to be supervised by adults. Take note: Some of the projects involve the use of matches, volatile materials, and projectiles. Adult supervision is mandatory for all such projects.

2. Read the entire project description carefully before beginning the project. Make sure you understand what the project is about and what you are trying to accomplish. If something is unclear, reread the directions until you fully comprehend the entire project.

3. Don't make substitutions for the specific liquids and aerosols indicated for use in each project. Stay away—far away—from gasoline, kerosene, starting ether, alcohol, and other flammable liquids. There are few things as dangerous as flammable liquids or aerosols. They can and do explode, and the consequences can be disastrous.

4. Use only the quantities of fluid listed in the project instructions. Do not use more propellant than specified.

5. Don't make substitutions in materials or alterations in construction techniques. If the directions say to cure a joint overnight, then cure it overnight. Don't take shortcuts.

6. Read and obey all label directions for materials such as PVC primer, cement, and any other chemicals. Pay close attention to directions regarding ventilation.

7. Remove and safely store all cans and bottles containing naphtha, hairspray, or any other flammable substance prior to launching the project or performing the experiment. A good rule of thumb is to maintain a hazard-free radius of at least 50 feet around the area in which you plan to work.

8. The area in which the projects are undertaken must be cleared of all items that can be damaged by projectiles, flying objects, and so forth.

9. Be aware of the hazards on projects with mechanical arms or other moving parts. Keep fingers and toes clear of pinch points.

10. Keep people away from the firing zone in front of and to the sides of rockets, mortars, cannons, etc. Use care when transporting, aiming, and firing, and always be aware of where the device is pointing.

11. Wear protective eyewear as you build and operate the projects described. Some experiments also call for hearing protection, blast shields, gloves, and so forth. Always use them.

REMEMBER THIS

The instructions and information are provided here for your use without any guarantee of safety. Each project has been extensively tested in a variety of conditions. But variations, mistakes, and unforeseen circumstances can and do occur. Therefore, all projects and experiments are performed at your own risk! If you don't take this seriously, then put this book down; it is not for you.

Finally, there is no substitute for your own common sense. If something doesn't seem right, stop and review what you are doing. You must take responsibility for your personal safety and the safety of others around you.

WORKING WITH PVC PIPE

Several of the projects in this book involve cutting and joining PVC pipe. PVC pipe is normally used for water pipes in homes and buildings and our use of it here is not how the manufacturer intended. But in my experience, it works adequately for the projects in this book, but of course, your experience may vary.

First, you should be aware that at least four types of plastic pipe and plastic pipe joints are available: PVC, CPVC, ABS, and PB. The letters are abbreviations for the type of plastic polymer that composes the pipe. Pressure-rated schedule 40 PVC pipe and pipe fittings are made of white polyvinyl chloride. This is the type of pipe and pipe joint recommended for most of these projects.

You can purchase the PVC pipe and pipe fittings at your local hardware or home store. Note that even among PVC pipe there are different kinds, and some, such as cellular core PVC, are not suitable for use in these projects. **Be sure to use only pressure-rated schedule 40 PVC pipe with the pressure rating printed on the outside of the pipe by the manufacturer.** If there is no pressure rating or schedule number printed on the pipe, don't use it. Such pipes are likely made from cellular core materials, which are not strong enough.

The pressure rating of the pipe changes with temperature. As the air temperature rises, the plastic gets weaker. At 90°F, the pipe's pressure rating is only three-quarters of what it is at 70°F, and at 100°F, the pipe is only two-thirds as strong. While projects in *Whoosh Boom Splat* are designed so pressures are always well below ratings, keep in mind that on hot days your margin of safety is much less.

Unlike pipes, PVC fittings (elbow joints, tee joints, and so forth) usually do not come with pressure ratings on them. Typically they are rated for about 150 psi, although some are rated for more and a few for less. If you are in doubt of the pressure rating of the fitting, ask the person you are buying it from. Be sure to inspect the pipe end and fittings for cracks, dirt, and abrasions. Don't use damaged PVC pipe or fittings.

Cutting and Fitting PVC Pipe

PVC pipe is easily cut with a regular handsaw. It is important that you make all cuts as close to 90 degrees to the centerline of the pipe as possible. That way, you won't leave any interior gaps, which will weaken the joint. Be sure to remove any burrs.

You may want to dry fit the pipe into the joints before you apply any cement, to see how things fit. Sometimes the dry-fitted pipes and joint fittings stick together so tightly it is hard to get them apart. If that happens, you can carefully whack the fitting loose with a wooden block.

Joining and Cementing PVC Pipe

The process of joining and cementing PVC pipe is technically called "solvent-welding." The solvent melts the plastic so that when you push the pipe and the pipe fitting together, the two parts fuse as the solvent evaporates. Each type of plastic pipe has its own special solvent. Some solvents are advertised to work on several types of plastic, but it is recommended that you use the solvent that is meant solely for the type of plastic you're working with. At the hardware store the solvent you need is called "PVC cement."

The solvent works only on *clean* surfaces—surfaces with no dirt, no grease, and no moisture. Wipe the inside of the fitting and the outside of the pipe with a clean cloth. Then apply PVC primer (called "purple primer") to the ends. Allow the primer to dry before applying cement.

Next, coat the surfaces that you want to join with a thick coat of PVC cement. Apply plenty of cement, first to the pipe end, then to the fitting socket. Leave no bare spots. PVC cement, which is a solvent, should be used in well-ventilated areas. Immediately join the pipe and fitting full depth with a slight twist to bring it into correct alignment. A continuous ooze of cement around the fitting indicates that you used enough solvent cement to ensure a leak-free joint. Let the joint dry for several hours before using. Be sure to observe the cure time shown on the PVC cement can's directions. The table on page 12 provides average joint cure times.

Once joined, you can't reposition the pipes or otherwise fix errors. If you accidentally put the wrong fitting on a pipe, you have no choice but to cut it out and start over.

GENERAL PVC SOLVENT SAFETY INSTRUCTIONS

1. Avoid breathing the PVC solvent cement and primer vapors. Work in a well-ventilated area.
2. PVC solvent vapors *are flammable*. Keep all chemicals away from open flames and observe cure times carefully before using.
3. Read and follow the precautions that appear on the labels.
4. Joint Cure Time is the time required before pressure testing. In damp or humid weather, allow 50 percent additional cure time. How long should PVC cure? Use the table here.

Average Joint Cure Times for PVC Solvent Cements

Temperature during assembly or cure period	Pipe Diameter ½" to 1¼"	Pipe Diameter 1½" to 3"
60°–100°F	6 Hours	12 Hours
40°–60°F	12 Hours	24 Hours

EARLY ARTILLERY AND THE NIGHT LIGHTER 36

Prior to 399 BC there was no such thing as artillery. Ancient warfare consisted of hand-to-hand combat—club against club or sword against sword. In addition, warriors used thrown weapons like spears and stones, which were usually hurled from slings. Finally, they shot arrows from stretched bows. In 399 BC military leaders in the Greek city of Syracuse came up with ideas for larger, semistationary weapons that could wield much more power than could a single warrior's arm. These devices used animal parts, such as horns or neck tendons or hair, which had been fashioned into springs. By stretching these springs, a man (or even better, a group of men) could load the device with comparatively huge amounts of potential energy. These weapons were the earliest catapults and represented the first type of artillery.

From 399 BC until about AD 1330 all military artillery was powered by human or animal muscle. During this 1,730-year stretch, artillery took the form of siege weapons called onagers, trebuchets, ballistas, mangonels, petraries, and spring engines—all of which we now refer to by the general term *catapult*. Catapults work by suddenly releasing energy that has been loaded into the device by work from the operators' muscles. In ancient and medieval times catapult operators would tighten a rope spring, bend back an enormous bow made from wood and animal horn, or raise a great weight high off the ground. When a catapult is fired, a spring releases or a counterweight falls and, through a mechanism or lever, quickly and efficiently transfers its stored energy to the projectile.

Catapulting a bigger stone naturally requires a larger counterweight or a thicker composite bow. By the end of the 13th century, catapults were getting about as big as they could get. The law of diminishing returns was taking over; making catapults any larger and more powerful would push the materials from which they were made to their mechanical limits. Moving and erecting the largest catapults already required gigantic siege trains and armies of men. An entire forest had to be cut to provide wood for such a device. Given the deforestation throughout Europe at the time, building such giant catapults became an ecological disaster as well as a logistical nightmare. Most important to kings and princes, they simply became an unacceptably large drain upon royal treasuries.

Such was the state of artillery warfare throughout the Middle Ages. But then a new discovery, brought by explorers and traders returning from the Far East, changed the world. The Chinese called this discovery the "fire drug." We know it today as gunpowder.

Fire-based, or incendiary, projectiles had been used in warfare dating back to the ancient Greek and Chinese cultures. Ancient accounts tell of baskets of flaming oil being hurled at enemies, arrows tipped with heads of burning straw, and even the Fire Ox, a Chinese innovation that is just what it sounds like—an ox with a barrel of flaming goo strapped to its hindquarters and sent stampeding toward enemy lines. But all of these were weapons of conflagration, not explosion. They spread fire, but they did not blow up with great and destructive energy.

Around AD 1000, the Sung dynasty in China began experimenting with the "fire drug," a compound of charcoal, sulfur, and a white powder that leached to the surface of the ground in certain areas.

The white powder was potassium nitrate, which is often referred to as saltpeter.

By themselves, charcoal and sulfur do a pretty good job of burning. They release energy when ignited, but they do it rather slowly. The Chinese discovered that adding saltpeter to a mixture of sulfur and charcoal caused the compound to burn far more rapidly. And if the ratio of charcoal, sulfur, and saltpeter was exactly right, the compound would explode when heated in an enclosed space.

The Chinese military quickly made use of this discovery. They placed the fire drug mixture in various types of projectiles and bombs, giving them elaborate names such as "Bandit-Burning Vision-Confusing Magic Fire Ball," "Match-for-Ten-Thousand-Enemies Bomb," and "Heaven-Shaking Thunder Crash Bomb."

The Chinese certainly knew this mixture was very powerful. Scientists now know that power is due to gunpowder having a fuel-oxidizer chemical combination. Things burn because they react with oxygen, and the more oxygen available, the better things burn. Saltpeter is an efficient oxidizer. When it is burned, a chemical reaction releases a great number of oxygen molecules that combine with the burning sulfur-charcoal fuel. The result is an ultra-vigorous energy release. When this reaction occurs in a closed container, an explosion results, creating a weapon worthy of the appellation "The Bone Burning and Bruising Magic Oil and Smoke Bomb."

Ancient chemists tried different proportions of charcoal, sulfur, and saltpeter and found that minute proportional changes resulted in much improved results. A bit more saltpeter and a bit less sulfur here and there would substantially increase the power of the chemical compound. Over time the experimenters found that the most effective proportion was about 75 percent saltpeter, 15 percent charcoal, and 10 percent sulfur. Once the correct proportions of the "devil's distillate" were determined, military men began to consider how to use it. One of the most straightforward applications was as a propellant for cannonballs.

One of the first trustworthy reports of the use of a gunpowder cannon (or at least primitive cannonlike technology such as a serpentine or culverin) in battle occurred in 1324. In the 14th century the city of Metz had become one of the largest in eastern France because of its position on the intersection of important trade routes. Rivaling some of the great cities of Flanders, Italy, and Germany, Metz was a tempting target for power-hungry noblemen. In 1324 the Dukes of Luxembourg and Lorraine attempted to take the city by force. According to the old chronicles, one of the leaders of the siege, William de Verey, sailed into the town onboard a flatboat, loaded not with horsemen or catapults but with new and secret weapons: things that had not been seen in warfare previously but that would be used in virtually all battles from that time onward.

An account written in 1838 by French historian J. Huguenin in *Les Chroniques de la Ville de Metz* relates that on this barge were "*serpentines et canons . . . et tirant plesieurs coptz de' artillerie*" (serpentines, cannons, and many other types of shooting artillery). According to this town history, it was here, at the walls of Metz, that European besiegers first packed the new explosive called gunpowder into iron vases.

The chroniclers of the battle don't tell us what effect, if any, the primitive rock-firing cannons had on the outcome of the siege. Most likely, despite the presence of gunpowder cannons, the battle was won by the tried and true methods of medieval siege tactics: escalade (scaling walls with ladders), sapping (digging tunnels), frontal assaults, and subterfuge. But in a short time, artillery improved to the point where such medieval siege tactics were no longer relevant. Modern warfare probably got its start there, on a river barge filled with crude cannons that literally could not hit the broad side of a barn.

HIGH VOLTAGE AND NIKOLA TESLA

Our first project combines mechanical principles of explosive artillery warfare with the technology of electricity to produce a unique and powerful handheld spud gun. The Night Lighter 36 is a combustion-based cannon capable of shooting a potato from one end of a football field to the other. Its power and efficiency result from a high-voltage spark, thanks to the advances of electrical engineering. So it's important to look back at Nikola Tesla, who brought high voltage into daily life in a meaningful way.

Nikola Tesla was born in 1856 in Smiljan, Croatia. As a child, all who knew him realized that he was in possession of incredible intellect but was simultaneously plagued by myriad mental phobias and compulsions.

Tesla attended several schools and colleges, most notably the University of Prague, until he was banished for excessive gambling and "leading an irregular life." While in Prague, Tesla started work on what would be his most important contribution to science: alternating current, or AC.

Prior to Tesla's work, electrical power generation systems were limited to producing low voltages and high currents. Power companies had to either use very expensive, heavy copper transmission cables (to minimize wire resistance) or build a direct current generating station every mile or two. This low-voltage method, known as DC power distribution, was favored by the enormously rich and powerful American businessman Thomas Edison.

But high voltage, Tesla felt, was the key to the cheap and ubiquitous distribution of power, and using alternating current was the only way to achieve such distribution. He found a champion in a wealthy businessman named George Westinghouse. A brilliant and far-sighted entrepreneur, Westinghouse had built a giant business enterprise from nothing, making his fortune by manufacturing air brake systems for trains.

The new Westinghouse Electric Company was well financed, well managed, and most important, willing to base its entire business plan on Tesla's idea of AC power, which in almost every way was superior to DC. But Edison would not give up. Even if AC power was exponentially more efficient to transport and cheaper to make, it did have liabilities, and these he would exploit with a Madison Avenue–style public relations appeal to the public.

The Edison camp's plan was simply to portray AC power as too dangerous and deadly for use in homes and businesses. The PR campaign went something like this: AC power uses high voltage to transport power into people's homes. In fact, high voltages in and of themselves make the whole concept fly. But, said Edison's PR minions, high voltages are dangerous. If you accidentally touch high voltage, you can die. Today every room in every home and business has several 110-volt AC outlets, but in 1883 the idea of loading up your house with portals to such deadly currents was nearly unthinkable.

"AC power can kill you and your family," the Edison PR flacks said, and they backed up their claims with dubious tactics and statistics. Worse, they graphically demonstrated their point by electrocuting (with AC power, of course) scores of animals, barnstorming the country in a ghastly road show. The Edisonians fried scores of stray dogs and cats as well as farm animals (and in Albany, New York, an orangutan). The animal-electrocution cavalcade reached its cynosure with the Coney Island electrocution of a rogue elephant named Topsy.

But despite Topsy and the rest of the PR campaign, Edison's hope of discrediting Westinghouse failed, and soon Westinghouse's AC power was lighting the incandescent lamps of all electrified localities. Tesla's higher-voltage AC model won out: it was cheaper, faster, and just plain better.

In 1896 Tesla successfully installed a large AC power system at Niagara Falls, New York. The completion of that system ended the "War of the Currents," as this technology competition is popularly called. Soon afterward Edison's General Electric Company capitulated and converted to AC power.

Interestingly, New York City's electric utility company, Consolidated Edison, continued until recently to supply DC current to a few customers who had specialized needs for it, mostly for old elevators. Only in January 2005 did Con Ed announce that it would cease offering DC service to those 1,500 users, ending that chapter of electrical transmission history.

Bad Elephant Killed

Topsy, the ill-tempered Coney Island elephant, was put to death in Luna Park, Coney Island, yesterday afternoon. The execution was witnessed by 1,500 or more curious persons, who went down to the island to see the end of the huge beast, to whom they had fed peanuts and cakes in summers that are gone. In order to make Topsy's execution quick and sure 460 grams of cyanide of potassium were fed to her in carrots. Then a hawser was put around her neck and one end attached to a donkey engine and the other to a post. Next wooden sandals lined with copper were attached to her feet. These electrodes were connected by copper wire with the Edison electric light plant and a current of 6,600 volts was sent through her body. The big beast died without a trumpet or a groan.

Topsy was brought to this country twenty-eight years ago by the Forepaugh Circus, and has been exhibited throughout the United States. She was ten feet high and 19 feet 11 inches in length. Topsy developed a bad temper two years ago and killed two keepers in Texas. Last spring, when the Forepaugh show was in Brooklyn, J. F. Blount, a keeper, tried to feed a lighted cigarette to her. She picked him up with her trunk and dashed him to the ground, killing him instantly.

From the Commercial Advertiser *(New York), January 5, 1903*

The Night Lighter 36

Spud guns are very popular amateur science projects. Few devices can be made as simply and inexpensively as a spud gun yet provide so much entertainment and fun. This project shows you how to make the Night Lighter 36, a transparent PVC potato cannon with a stun gun ignition system. It can be used both day and night, but the clear PVC tubing provides an excellent view of the internal workings when used in darkness.

The know-how for fabricating potato cannons is provided in many books—including my own *Backyard Ballistics*, which includes detailed instructions for making a simple yet dependable one—as well as on various websites. There are companies that sell the plans,

Author with Night Lighter 36

parts, or completed spud guns over the Internet. Some products feature rather elaborate construction, including rifled barrels and metered fuel systems.

Most spud guns, however, are made at home. Having once mastered basic gun construction, the intrepid potato cannoneer often wants to design and assemble more complex and more artistic devices. The word *artistic* is applicable because making spud guns is an art as well as a science. For many, the act of shooting resonates on a primal level, and makers often take great pains to enhance their experience by continuously tinkering with their guns with the ardor of hot-rod builders and model railroaders.

The Night Lighter 36 potato cannon uses inexpensive modern circuitry to produce a stunningly high voltage that ignites a fuel-air mixture in the cannon barrel. It is a relatively simple extension of the basic spud gun concept and is fun to operate and entertaining to behold. A basic, no-frills spud gun can be built for less than $25. The Night Lighter 36 costs more due to the price of the transparent PVC tubes and the stun gun igniter. But with a bit of scrounging, you can often control costs. I built mine for less than $50 by finding odd pieces and leftovers from plastics suppliers. If the cost of the transparent PVC pipe is a problem, you can build this project using regular, nontransparent, schedule 40 PVC pipe.

The Grandfather of the Spud Gun

The hard white plastic used to build spud guns is called PVC, an abbreviation for polyvinyl chloride. It was invented in 1926 by a scientist named Waldo Semon. At the time Semon was a young research scientist working for the B. F. Goodrich Company in Akron, Ohio.

Semon accidentally discovered PVC while he was attempting to develop synthetic rubber. In World War I, Germany's adversaries cut off its supply of natural rubber. In response, the Germans developed rubber substitutes. Their efforts were only partially successful at best—the substitutes were not very resilient. Drivers of cars with synthetic rubber wheels would often find that the tires had formed flat spots overnight, resulting in a bumpy ride the next morning. Worse, the synthetic rubber was more expensive than the real thing.

Shortly after the end of World War I, Goodrich determined that a low-cost, high-performance rubber substitute would be a product with enormous economic potential. Since Semon was one of their best scientists, company management instructed him to begin work to find a better-behaved, less expensive man-made rubber.

During his extensive research, Semon came upon the work of earlier scientists who had developed a substance called "oil of Dutch chemists." Another chemist, he found, had created a gas called vinyl chloride from the Dutch oil. Semon felt the stuff had potential, and so he started experimenting with vinyl polymers. Polymers are organic substances with very large molecular chains. After several lab explosions and near misses, his chemical experimentation yielded a hot powder that, when cooled, turned into a flexible but tough gel. This was the first batch of polyvinyl chloride.

Interestingly, Semon took the most pleasure in another of his inventions that was far less heralded—synthetic rubber bubble gum. "It looked just like ordinary gum, except that it would blow these great big bubbles," he said. To his disappointment, it never caught on.

Potato cannons and their relations have been built from pipes made of steel or aluminum, or other types of plastic, or even reinforced fabric tubing. But for cost, strength, and ease of fabrication, PVC is hands down the most popular material. While Dr. Semon might never have made his own spud gun, his invention made them possible.

Safety

If you want to join the community of spud gun enthusiasts, you must understand a couple of important points.

The first is that spud guns may not be legal in your area. That may seem unfair, considering that in some places it is perfectly legal to walk down Main Street holding a twelve-gauge, but toting a tater gun may get you in trouble. While no federal law appears to regulate the construction and use of potato cannons, some state and local laws do. So check with local law enforcement regarding the rules that pertain to your area, then obey them.

Second, spud guns can be dangerous—they can shoot with great power and force. So treat your gun with respect. A few important points to keep in mind:

1. A potato cannon shoots with enough force to cause injury. Its range can exceed 200 yards. Always use extreme care when handling and aiming the device. Wear eye protection.
2. Never, ever look down the barrel of a gun.
3. Don't operate a damaged or worn gun. Check the cannon frequently for signs of wear.
4. Don't take construction shortcuts, use inferior materials, or make substitutions in the materials list. But if clear PVC is too expensive for your budget, you may substitute white schedule 40 PVC pipe.
5. Review the guidelines on working with PVC on pages 9–12. PVC's mechanical properties degrade in high and low temperatures. Don't use spud guns in temperatures below 55° or above 95° F.
6. The vapors from PVC cement are flammable. Allow all joints to dry fully before exposing them to ignition sources.
7. A misfire is a fairly rare event, but if it does occur, use the utmost caution. Carefully remove the end cap and ventilate the combustion chamber thoroughly. Never look down the barrel, and never point the barrel at anything you don't want to hit.
8. The stun gun igniter is made from an inexpensive stun gun commonly used for personal protection. Check local laws regarding their possession and use. Stun guns are designed to be high-voltage shocking devices. Touching

the high-voltage electricity contacts of a stun gun will hurt. Avoid them!

9. As with many popular activities, things occasionally go wrong. If you follow the directions here carefully, the chance of a bad outcome is remote, but unexpected things can happen: material flaws, bad judgment, pure bad luck. Always use common sense, and take careful note of the following: *Build and use the spud gun at your own risk.*

TOOLS

- Hacksaw
- Power drill
- Sanding drum for power drill
- $^{15}\!/_{64}$-inch drill bit
- Utility knife
- Screwdriver
- Pliers
- Wire stripper
- Safety glasses

MATERIALS

- (1) 14-inch-long, 3-inch-diameter transparent (or regular) schedule 40 PVC pipe
- (1) 36-inch long, 2-inch-diameter transparent (or regular) schedule 40 PVC pipe
- (2) Crimp-on barrel connectors
- (1) Stun gun, nominal 100,000-volt output or better
- (2) 2-inch-long, ¼-inch hex head bolts, full threaded
- (4) ¼-inch hex nuts
- (1) 3-to-2-inch-diameter white PVC reducing fitting
- (1) 3-inch white PVC threaded adapter coupling, one side smooth, one side threaded
- (1) Can PVC primer
- (1) Can PVC cement
- (1) 2-foot length of insulated 12-gauge wire

- (2) Large crimp-on spade connectors
- (2) 15-inch-long (not diameter) hose clamps
- (2) Small wire nuts
- Electrical tape
- (1) Tube of silicone sealant
- Bubble wrap (optional)
- Bag of potatoes
- (1) 4-foot long, 1-inch-diameter wooden dowel or broom handle
- (1) Can of hydrocarbon containing aerosol spray (hair sprays typically work well)
- (1) 3-inch-diameter white PVC threaded end cap

PREPARE YOUR MATERIALS

1. First, cut the PVC to size. Measure and mark a cutting line 14 inches from one end of the 3-inch-diameter PVC pipe. Use the hacksaw to cleanly and squarely cut the pipe.

1.1

2. Now, measure and mark a cutting line 36 inches from one end of the 2-inch-diameter PVC pipe. This will be the cannon's barrel.

1.2

3. Fillet one end of the barrel as shown in **diagram 1.2**. Use a drill and sanding drum attachment to taper one end of the 2-inch-diameter pipe section so it forms a sharp edge. A clean sharp edge is important since it should cut the perfect-sized potato plug projectile as you ram the potato into the muzzle of the gun.

1.3

4. Drill a hole 4 inches from one end of the 3-inch-diameter pipe using a $^{15}/_{64}$ inch drill bit. Drill a second hole of the same size, 4 inches from the same end, but at the spot that is exactly diametrically opposed to the first hole.

5. Now modify the barrel connectors. Using a sharp utility knife, remove excess insulation from the barrel connectors as shown in **diagram 1.4**. With the stun gun turned off (temporarily remove the battery for extra safety), place the connectors over the stun gun electrode to test the fit. Be sure to use the twin electrodes that point forward rather than toward each other.

1.4

Note: Depending on the make and model of the stun gun you use, you may need to modify these directions in order to connect the ignition wire to the stun gun electrode. Other types of connectors such as wire nuts and soldered connections may be used if necessary.

BUILD THE NIGHT LIGHTER 36

1.5

1.6

1. Install the electrodes. Screw in the 2-inch-long hex head bolts with two hex nuts attached on each (nuts go outside the barrel) into the holes in the 3-inch-diameter pipe. Allow the bolts to self-tap themselves into the softer plastic for a tight fit, but do not

overtighten them or you'll strip the PVC. Position and adjust the nuts as needed so there is a spark gap of approximately ¼ inch between the bolt ends inside the barrel as shown in **diagram 1.6**.

1.7

2. Solvent-weld the 3-to-2-inch reducing fitting to the end of the 3-inch pipe closest to the electrode bolts (see **diagram 1.7**). Clean the mating surfaces with PVC primer. Then attach the reducing fitting, carefully following the directions given on the PVC cement container (see also page 11–12).
3. Solvent-weld the smooth side of the 3-inch threaded adapter coupling to the other end of the 3-inch pipe.

1.8

4. Solvent-weld the 2-inch barrel to the corresponding opening in the 3-to-2-inch reducing fitting as shown in **diagram 1.8**.
5. Let the assembly dry overnight.

WIRE THE NIGHT LIGHTER 36

1. Cut the 12-gauge wire into two 1-foot-long pieces. These will become the ignition wires. Crimp the spade connectors to one end of each ignition

1.9

wire. Crimp the unmodified end of the barrel connector to the other end of the wire.

2. Remove the battery from the stun gun. Attach the stun gun body to the gun barrel using the hose clamps, but be careful not to overtighten them. Position the stun gun body 90 degrees to the axis of the electrode bolts (see **1.10**).

1.11

1.10

3. Attach the ignition wire spade connectors to the electrodes as shown in **diagram 1.11**.

4. Attach the ignition wires to the stun gun electrodes using the modified barrel connectors. Crimp them on carefully.

5. Cover the stun gun test leads (inboard electrodes) with wire nuts, which may need to be cut down in order to fit. See **diagram 1.12**.

1.12

1.13

6. Insulate all exposed metal areas of the ignition path on the stun gun and bolt electrodes with electrical tape or silicone sealant. It's easy for electricity to find its way underneath any insulation gap at the base of the electrodes. Cover the bolt connections with silicone sealant. You may want to wrap the stun gun body with bubble wrap, but the stun gun operates at such high voltage that the insulating wrap will not completely prevent shocks. Use caution to avoid contact between the stun gun's primary electrodes and your skin. Don't be the path of least resistance!

Leveling Your Weapon

As must have been apparent to any 14th-century artilleryman in the vanguard of an assault, a round stone or iron ball doesn't travel very far or very accurately. But bullet-shaped projectiles are a relatively new innovation. Until the mid 1800s, firearms were mostly smooth-bored, muzzle-loaded weapons, and most projectiles were round.

If the musketeer put enough gunpowder in the charge, he could impart a reasonable muzzle velocity when the projectile left the barrel. But air resistance acting on the round shape quickly slowed it down. Contrary to intuition, spheres, it turns out, are among the worst possible shapes in terms of creating drag. And cylinders aren't much better. The problem with spheres and cylinders is that the wake they leave behind as they fly through the air is huge in proportion to their cross-sectional area. Big wakes cause big drags.

It's easy to see why Turkish cannonballs weren't very aerodynamic

In the late 1980s Austrian researchers conducted a series of tests on the effect of air friction on bullets for a group of 16th-through-18th-century muskets. What they found was surprising. The tests showed that when a spherical bullet is first shot, it decelerates at a rate of about 2.5 meters per second for every meter of distance it travels. Modern bullet-shaped bullets lose only about 0.9 meters per second in velocity over similar distances. In other words, spherical bullets slow down three times as fast as bullet-shaped bullets. After 100 meters, a musket-fired ball has only half as much kinetic energy as when it first leaves the muzzle. Modern rifle-fired ammunition maintains 90 percent of its energy at the same distance. Why? A bullet's cylindrical body and pointed head cut down on air friction—that is, aerodynamic drag—as it flies through the air. And the cylindrical body is designed to spin as it travels up the helical groove of a rifle barrel. The spinning stabilizes the bullet as it's in flight, allowing it to travel farther and straighter.

In addition to velocity loss, round shot is hard to aim accurately. Instead of flying straight and true toward the target, a round bullet tends to slice and hook like a golf ball. This effect is caused by an unintended spin that is imparted to the bullet around an arbitrary axis. The spin changes the air pressure on one side or the other of the sphere and sends the projectile off course. This phenomenon is called the Magnus effect, named after Gustav Magnus, who first described it in 1852. The veering of projectiles caused by the Magnus effect made it nearly impossible for a musketeer to hit a target.

So as late as the Battle of Waterloo in 1815, soldiers never talked about "aiming" their weapons. Instead, they merely "leveled" them.

OPERATING THE NIGHT LIGHTER 36

1. Carefully push a potato into the cannon from the muzzle end. The sharp cutting edge ground into the muzzle will cut the potato into a plug of the correct size. The potato must fit snugly on all sides of the muzzle. Any gap will allow the expanding gas to blow by the potato. If that happens, the potato won't fly far.
2. Use the dowel or broom handle to push the potato plug 30 inches down into the cannon muzzle.
3. Put the battery back into the stun gun. Put on your safety glasses. Direct a stream of aerosol deodorant or hair spray (check the label to make sure it contains hydrocarbons such as alcohol, propane, or butane) into the firing chamber (the 3-inch-diameter cylinder where the electrodes are). Start out with a 1-to-2-second burst. You'll likely need to determine the best amount of propellant by trial and error. Warning: The fluid stays in the combustion chamber and evaporates rather slowly. Always treat the gun as if there were flammable vapor in the firing chamber.
4. Immediately screw the end cap on securely.
5. Turn the stun gun on. Observe all safety rules.
6. Press the ignition button. Enjoy your work!
7. Many aerosol propellants contain chemicals that tend to gum up the inside of the cannon. Clean out the gun every few shots with a rag and nonvolatile cleaner (such as 409 or Simple Green). The end cap can get stuck from the gum. Use a wrench to unscrew it if necessary.

THE JAM JAR JET

The first successful cruise missile, the Nazi-built V-l *(V* stood for *Vergeltungswaffen,* which translates to "weapon of revenge"), was built by Germany during the middle years of World War II. The British called it the buzz bomb or the doodlebug because of the noise it made as it flew toward them. The engine that powered the V-l wasn't a piston engine, and it wasn't a rocket motor. It was a unique type of jet engine called a pulse jet.

The pulse jet engine is simple, cheap, and powerful but isn't used in commercial aviation because it is incredibly noisy, burns through fuel like a forest fire, and vibrates like a gigantic, unbalanced chain saw. Pulse jets were invented in Europe in the early 20th century but really had no practical use until the German scientist Paul Schmidt devised a dependable system of valves and fuel-delivery components for the Luftwaffe's terror weapon.

The V-l flying bomb is an example of what's now called a cruise missile. A cruise missile is much like an automatically guided airplane: it has wings, a tail, and a motor. It is an "air-breather"; that is, it uses a fuel-air combustion cycle engine for power, not a chemical rocket engine.

Today the US Air Force uses cruise missiles such as the Tomahawk, the ACM, and the Harpoon. Many countries build cruise missiles. For example, the Indians make the Brahmos and the Chinese have the Silkworm. Cruise missiles are a popular armament because they are relatively easy to build.

The German V-l was a tube about 25 feet long with two stubby wings extending from the fuselage. Carried piggyback on the rear was the pulse jet engine. The V-l had a range of about 150 miles and could carry an explosive payload of nearly a ton.

THE V-1'S TEST PILOT

For the era in which it was constructed, the V-1 was a highly technical and difficult-to-engineer weapon. Much was promised, but time and again during its development the V-1 (officially called the Fi-103 by the Nazis) disappointed its designers by crashing during or just after takeoff. According to the engineering analysis, everything looked fine on paper. On the ground, all systems tested perfectly. But as soon as the Fi-103 roared down the ramp and took off, it lurched out of control, sending the engineers and their angry bosses diving for cover. The only way to figure out what was going on, it appeared, was to put a live test pilot onboard to observe.

Only a special person is willing to strap into a tiny cockpit on a jet-powered missile that has a propensity for crashing in a ball of flame. The Nazis would have to find a brave man indeed to attempt something so dangerous.

No man came forward. But Hanna Reitsch did. Reitsch was one of Germany's best, and bravest, test pilots. She was petite, young, and brave—and had endured numerous crashes. She test-piloted the Messerschmitt 163B, a dangerous and ultrafast rocket-powered airplane. She also flew the FW-61, the world's first helicopter, indoors at an auto show in Berlin in 1938.

An expert flyer, she was also, unfortunately, an unrepentant Nazi. She was a confidante of Himmler, Goebbels, and Hitler himself and was one of the last people to see Hitler alive in his Berlin bunker just before the Red Army stormed in.

Why this woman would so wholly subscribe to such a hateful ideology is a mystery. More evidence of her perverse thinking is that toward the end of the war, she promoted the idea of airplane squadrons composed of "suicide pilots." The idea was to have fanatical German pilots fly modified V-1s into high-value targets such as Allied ships and headquarters buildings. Although the suicide squadron idea was never adopted by the German Luftwaffe, the concept was later embraced by the Japanese kamikaze corps.

At the German test site of Peenemünde, Flugkapitan (Captain of the Air) Reitsch strapped herself into the tiny payload section of a modified V-l and flew four hazardous trips in the flying bomb. Three times she took off and could not find the answer to the V-l's problem. On the fourth try she determined that the bolts holding the wing to the V-l's body were breaking off when the pulse jet kicked in, causing the V-l to become unstable and fly off course.

With the design problems fixed, the Germans cranked up production of the V-l using forced labor from concentration camps. The Nazis aimed thousands of buzz bombs at London from launching areas in France, Holland, and other nearby countries. A sensor on the front of the V-l determined roughly how far it had traveled since launch. When the preset distance was reached, fuel to the engine shut off and the V-l soared down toward its target in a deadly, silent glide. The sound of the buzz bomb wasn't what scared people. It was hearing it stop.

The first one hit London on June 12, 1943. At the height of their use, a few hundred were launched daily. But the British quickly became experts at intercepting the drones with Royal Air Force fighters and taking down many more with antiaircraft artillery. Consequently, only 25 percent of all V-1s launched ever reached their targets. The attacks ended when the Allies began to recapture land in occupied Europe, seizing launch sites as they went.

BUILDING YOUR OWN PULSE JET

You might believe that building a jet engine at home is beyond the means and capabilities of most hobbyists. But in this chapter, I'll show you how to build a simple jet engine—a pulse jet—for about $15 in materials and a couple of hours of labor.

Unlike, say, a turbojet or a fan-jet, both of which contain hundreds of precisely machined rotating parts, a person can build a working pulse jet in the garage with little more than hand tools and a welder. But even the need for those tools can dissuade some hobbyists. Those whose fabrication skills are limited to punching holes in sheet metal and bending wire can build a simplified version of the pulse jet: the Reynst combustor or, as it is better known, the Jam Jar Jet.

François Reynst was a Swiss engineer who experimented with a number of jet engine designs in the first half of the 20th century. As a youngster he showed keen interest in the science of combustion. One day while experimenting with alcohol in glass jars, he accidentally discovered an interesting phenomenon. He found that if he placed a small amount of alcohol in a closed jar and punched a hole of a certain dimension in the lid, a weird pulsating combustion occurred when the fuel was ignited.

Reynst looked on with great interest as flames shot out of the hole. The pulse jet combustion cycle then, almost magically, sucked the flame back into the bottle and ejected it again. The whole process would repeat until the fuel was expended. Reynst had a jar that was like the lungs of Saint George's dragon—it literally breathed fire. What young Reynst had discovered was a peculiar cyclical burner that would become known as the Reynst combustor: a simplified pulse jet engine.

Occurring within the jar is what scientists call systematic oscillation, which is caused by the burning of air and fuel and the geometry of the container. When the fuel and air inside the jar ignite, the Jam Jar Jet generates a burst of hot gas that raises the pressure inside the jar. The high pressure pushes the gas out of the hole in the jar in a strong pulse. But the exiting gas has mass and inertia, and when it leaves, too much of it exits. All of a sudden there's a vacuum inside the jar.

Nature, as the Dutch philosopher Spinoza said, abhors a vacuum. So fresh air rushes back in through the same hole to fill the void. The returning fresh air mixes with the residual alcohol still inside the jar. Since the inside is still very hot, it combusts again, and the hot gas is forced out again. The cycle repeats—explosion, thrust, vacuum—generating measurable amounts of thrust until the fuel is expended.

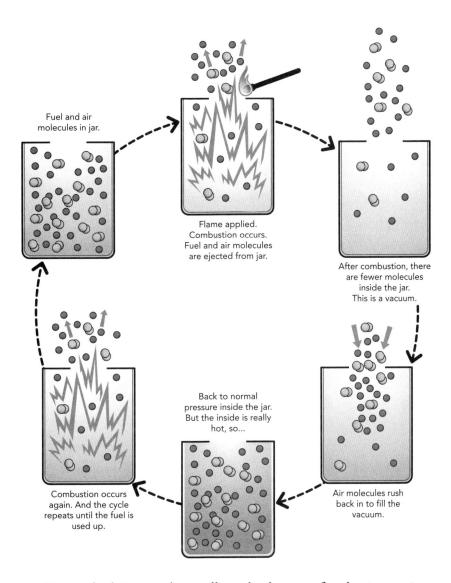

Fuel and air molecules in jar.

Flame applied. Combustion occurs. Fuel and air molecules are ejected from jar.

After combustion, there are fewer molecules inside the jar. This is a vacuum.

Air molecules rush back in to fill the vacuum.

Back to normal pressure inside the jar. But the inside is really hot, so...

Combustion occurs again. And the cycle repeats until the fuel is used up.

Reynst had invented a really radical type of pulse jet engine. Unlike any other, it has only one opening. But for a jet like this, that's all you really need. Continuous combustion cycles, such as those that that occur within a turbojet, require that the cycle's intake and exhaust be separate. But with pulsating combustion, the separation is not necessary, as the intake and exhaust parts of the cycle are not simultaneous. The combustion chamber "breathes" in and then out through the single port.

No two ways about it, this simple little contrivance, which can be constructed in less than a couple hours on a kitchen table, is a type of pulse jet engine, and it operates on a real combustion cycle.

The Jam Jar Jet

The Jam Jar Jet consists of a pint-size mason jar with a roughly conical, copper air-shaping ring suspended from the lid by wires. The lid has a round, carefully drilled half-inch-diameter hole in its center.

The jar itself is the see-through combustion chamber; the half-inch-diameter hole is the entrance and exhaust port. The conical copper ring serves a couple purposes. First, its shape speeds up the exhaust stream and slows down the intake phase, providing better cycle efficiency. Second, the thermal mass of the copper absorbs heat, taking some of the thermal expansion strain off the glass walls of the jar.

Safety

This is a jet engine you are building—a tempest in a teapot, so to speak. I've never had any problems with this design, but because you're building it at home with parts you've procured yourself, no one can guarantee your safety. No one but you is responsible for your work. That said and understood, go forward with the project if you desire. If you do, here are the measures you should take to create a safe environment.

1. Do not experiment with different-sized jars and openings. Using too large a jar with too small an opening may result in an explosion rather than a combustion cycle.
2. Use neither more nor less fuel than directed.
3. Use only the parts listed in the directions. People often ask me whether they can substitute something else for a part, or if they can make something a different size. My answer is always no, since I've not tried it or performed any analysis on that substitution.

4. Protect yourself by wearing safety glasses, gloves, and a lab coat.

5. The Reynst combustor is an extremely efficient heating device. After just a few seconds of run time, the outside of the jar gets hot enough to burn skin. Do not touch the jar for five minutes after a successful run, and even then be sure to test it to make sure it is cool enough to handle.

6. The glass jar typically cracks after several successful runs. Once your jar cracks, carefully discard it by sweeping the entire assembly into a bag without touching it. Seal the bag and throw it in the trash. Jars are cheap enough, so just go get another one.

7. Keep spectators at a safe distance, at least 10 feet away.

8. Always use a long-handled match or barbecue lighter to ignite the engine. After ignition, there is an immediate pulse of hot gas. A cigarette lighter or short match is not suitable.

9. Examine all parts before and after ignition for wear, and discard any worn parts.

10. Always use common sense before, during, and after running the Jam Jar Jet.

TOOLS

- Power drill
- ½-inch drill bit
- File or sandpaper
- ⅛-inch drill bit
- Metal-jawed vise
- Wire cutter
- Cookie sheet
- Eyedropper or measuring spoon
- Safety glasses
- Gloves
- Lab coat

MATERIALS

- (1) Pint-sized mason jar with extra screw caps and lids
- (1) Conical air diffuser (See step 3) made from copper pipe reducers.

- (1) 1½-inch-to-1¼-inch copper reducing fitting*
- (1) 1¼-inch-to-1-inch copper reducing fitting*
- (1) 12-inch length of 22-, 24-, or 26-gauge magnet wire
- (1) Small bottle of methanol. The easiest place to buy a small bottle of fairly pure methanol is the local auto supply store. Many varieties of gas line antifreeze (e.g., brand names Heet and Pyroil) are mostly methanol. Check the label for ingredients before purchasing.
- (1) Package of extra-long fireplace matches or a long-handled barbecue lighter

OPTIONAL ITEMS

- Table salt
- Boric acid crystals
- Freezer
- (1) 1-foot-long, 1-inch-diameter plastic or iron pipe
- Pliers and heat-resistant gloves.

BUILD THE JAM JAR JET

1. Drill a ½-inch-diameter hole in the mason jar's lid. Use a file or sandpaper to remove the burr. If the hole is jagged and cannot be made smooth and round, discard the lid and try again with a new lid.

2.1

---

* Most copper pipe fittings can be found at a hardware store. These exact fittings may be hard to find. You may substitute combinations of other fittings to form a conical reducer of about the same size.

2.2

2. Drill four ⅛-inch-diameter holes in the smaller copper reducing fitting as shown in **diagram 2.2.** The holes should be located about ¼-inch down from the edge of the small end of the adapter. Space the holes evenly around the perimeter at 90, 180, and 270 degrees from the first hole.

2.3

3. Insert the large end of the small copper reducing fitting into the small end of the larger fitting (**2.3**). Press-fit them together firmly. This forms the conical air diffuser and heat sink. The total height of the diffuser should be less than half the height of the mason jar.

4. Cut four 3-inch-long wires from thin-gauge magnet wire. Loop a wire through

2.4

each one of the holes in the smaller copper reducing fitting, and tie them on. Extend the other end of the wires outward, radially, from the diffuser cone as shown in **diagram 2.4**.

2.5

5. Center the conical air diffuser in the middle of the jar. Crimp the wires over the edge of the jar so that the top of the copper cone remains suspended close to the top of the jar, with a gap of ¼ to ½ inch from the jar's lid.

6. Use an eyedropper or teaspoon to insert 5 milliliters (slightly less than a full teaspoon) of methanol into the bottom of the jar. The methanol should barely cover the bottom of the jar—you must use only a small amount.

2.6

7. Screw the mason jar lid down onto the jar and the copper wires. The lid will hold the copper cone securely in place at the top of the jar.

8. **Optional**: Place the Jam Jar Jet in the freezer for two minutes. Based on my experience, slightly cooling the fuel and the jar seems to improve performance.

2.7

9. Hold your thumb over the ½-inch opening in the lid. Swirl or shake the methanol inside the jar. Place the cooled Jam Jar Jet on a cookie sheet, and place the cookie sheet on a secure surface, away from flammable objects. When you remove your finger from

2.8

the hole, you should notice a slight pressure release, and the jar will make a very faint *pffft* sound. If there is no slight pressure and no sound, swirl the fuel again. If there is still no pressure,

there is a leak in the seal of the jar that must be fixed.

10. Wear safety glasses, gloves, and a lab coat. Using a long-handled match or a barbecue lighter, hold a flame over the ½-inch opening.
    The flame will ignite the fuel, and

2.9

immediately the Jam Jar Jet will fire up. For the next 5 to 15 seconds, the Jam Jar Jet will cycle, pulse, and buzz, operating on a low but audible frequency. With the lights down low, you'll enjoy a blue pulse of flame growing and shrinking under the lid as the jar breathes fire. The size and frequency of the jet pulses will vary according to conditions in the jar and in the atmosphere.

INTERESTING VARIATIONS

- Add a pinch of table salt to the methanol. This will cause the pulse to be bright yellow instead of intense blue.
- Add a pinch of boric acid crystals. The crystals will result in a deep green flame.
- Using pliers or a gloved hand, hold a tube ½ inch or so above the exit-entrance hole. Iron pipe or plastic works, and even a cardboard tube will last a little while. Experiment with the length and diameter of the tube. When the size is right, you'll be rewarded with a loud, deep, resonant buzz.
- Some Jam Jar Jet enthusiasts experiment with a metal jar instead of a mason jar, outfitting it with a resonant tube suspended above the hole. This is sometimes termed a snorkeler. Some snorkelers have fuel-feed systems that drip methanol into the combustion chamber and thus can sustain combustion for long periods of time.

PROBLEMS?

If the methanol ignites with a single big whoosh instead of pulsing:

- Check the size of the hole; make sure it is a full ½-inch diameter.
- Make sure the jar and methanol are not too warm. Put them in the freezer for a minute or two, and then try again.
- Make certain the jar is charged with the recommended amount and type of fuel.

If you hold the match over the opening, and it doesn't ignite:

- Methanol is hydrophilic and absorbs water readily from the air, so make sure the methanol is fresh. (The bottle's seal has not been removed for more than a day.)
- Cool down the jar in the freezer.
- Make sure you are using about a teaspoonful of fuel in the bottom, no more than that.
- Vigorously shake the fuel.
- Check the seal by listening for the *pffft* when you remove your finger from the hole. Rejigger the lid if necessary to get a good seal.

If the jar cracks:

- The Reynst combustor/pulse jet is a very efficient burner and therefore extracts a lot of heat from the fuel very quickly. If the jar you're using cannot handle the rapid expansion, it will crack. If this happens, carefully dispose of the broken jar and replace it with a jar with thicker glass if possible.

HIRED GUNS AND THE ELASTIC ZIP CANNON

Like the itinerant gunslingers of the old West, there have always been exceptionally clever scientists and engineers willing to sell their expertise in weapon-making to the highest bidder. This chapter features a mechanically powered project and also tells the tales of a few of history's most famous "soldier-engineers of fortune."

In the mid-15th century Europe was something of a mess. For the previous hundred years the French and English had engaged in a costly and bloody war that took thousands of lives and caused enormous suffering. The Black Death—bubonic plague—while perhaps not quite the scourge it had been a hundred years earlier, was still killing thousands in its frequent outbreaks. Perhaps nowhere else in Europe, though, was life as cruel as in Romania, where in 1456 Vlad III of Walachia ascended to power. Vlad was a minor prince in the Balkan region. But the flinty-eyed, black-mustachioed Vlad was a barbarian, a merciless and probably mentally ill despot worthy of keeping warped company with the likes of Caligula, Hitler, and Pol Pot.

British historian Robert Seton-Watson wrote that Vlad was "a man of diseased and abnormal tendencies, the victim of acute moral insanity." If contemporary reports are to be believed, he was a merciless despot who craved complete control and demanded complete subservience from his subjects. He was called Vlad the Impaler, Vlad Dracul, and Vlad the Devil, for he murdered thousands of people with little compunction and no mercy. His trademark act of barbarism was his preferred method of executing enemies, and the source of his nickname—impalement.

Vlad was a fierce fighter, and the soldiers his army captured usually regretted not fighting to the death. Vlad was said to enjoy eating dinner in his palace surrounded by the bodies of a host of newly impaled prisoners. Besides captured soldiers, Vlad impaled priests, merchants, nobles, bureaucrats, Transylvanians, Hungarians, Walachians, men, women, children—pretty much anybody he didn't like, which was just about everybody. According to one account, he once saw a monk riding by on a donkey and, angered by some minor transgression now lost to history, impaled them both on a single stake.

By 1461, Vlad had incurred the wrath of the powerful Ottoman sultan Mehmet II, known as Mehmet the Conqueror. Vlad had burned Ottoman villages and towns and impaled a reported 25,000 Ottoman civilians. Mehmet moved swiftly against the real-life model for Count Dracula (it was "Vlad Dracul" who inspired the vampire of Bram Stoker's novel).

Setting out to exterminate Vlad, Mehmet assembled more than 100,000 men and 200 ships, as well as his signature weapons, giant cannons, of which much more will be said shortly.

Vlad and his dark-hearted followers were hard to find and even harder to fight. They employed early guerrilla warfare tactics: frequent raids in which they materialized out of the woods, made a quick and deadly strike against the Ottoman Turks, and then disappeared. Vlad picked his targets carefully, often finding weak or poorly defended spots in the Turkish lines and inflicting great damage against Mehmet's much larger force.

Eventually the size and might of Mehmet's Turkish forces overpowered Vlad. The quantity of land troops, the ships, the cavalry, and the famed Turkish artillery cannonades drove the evil Vlad back to the remote mountains of Transylvania, where he continued to scourge his country.

After a couple years in Transylvania, the king of Hungary got his hands on Vlad and imprisoned him. Released from prison after just four years, he bided his time, hanging around royal courts for another 12 years under house arrest. His brother Radu "the Handsome" had become ruler of Transylvania. When Radu unexpectedly died, Vlad III made another play for power and invaded Walachia with a small army of supporters.

Mehmet remembered the destruction and ruin Vlad had unleashed just a decade earlier. Unwilling to allow Vlad to regain power, Mehmet sent an army of well-equipped Turks, with artillery and specialized fighters, deep into Romania. The Ottomans did what they set out to do: Vlad was killed, perhaps in battle, perhaps by his own troops. In any event, his head was separated from the rest of his body, placed in a jar of honey, and sent to Istanbul as proof to Mehmet that Vlad Dracul—Vlad the Devil—was finally dead.

THE WALLS OF CONSTANTINOPLE

In most accounts of Turkish military history from Mehmet's time onward, cannons and other artillery play an important part. There is probably no battle so important to Ottoman history and no battle in which cannons played so dominant a role as the taking of the Byzantine capital of Constantinople.

In 1453 Constantinople was known as the Queen of Cities, the ancient capital of the Byzantines, and the place from which Emperor Constantine XI ruled a receding but still important empire. Despite the decline, the Byzantines continued to take pride in their huge and beautiful public works, monuments, and statuary. For centuries Constantinople was the largest city in the Western world. Unsurpassed in wealth and power, it was considered nearly impregnable. That reputation turned out to be its downfall. Its richness and apparent invincibility made it a target for the ambitious Mehmet II.

The conqueror cared little about the theological value of the city's vaunted collection of religious icons and relics. But many were made of solid gold and silver and were festooned with huge gemstones and other valuables, making the city tempting for pecuniary reasons. Yet the city was surrounded by the legendary "Walls of Constantinople," a heavily fortified network of walls, earthworks, moats, and other barriers. Construction had begun in the fifth century, and for a thousand years the thick, fortified stone walls protected the city from attack. No city on Earth was besieged more often (by the Huns, the Rus, the Turks, the Bulgars, and the Arabs to name but a few failed armies) than Constantinople and with less success. Those walls had proven themselves unbreachable time and again, an impenetrable bastion against usurpers and conquerors.

Walls or no walls, in 1453 Mehmet determined that Constantinople would be his. He gathered his huge army and went west to lay siege to the Queen of Cities.

Around the mid-15th century siege warfare was being revolutionized by the invention of cannon. In the 1420s, after the Byzantines witnessed besiegers using crude cannons against them, they reduced the effectiveness of the new weapon by suspending bales of material, wood, and anything else that might absorb and diffuse a projectile's impact. But Constantinople's fifth-century fortifications were designed for spear-wielding infantry assaults supported by the occasional catapult barrage. They were no match for the new, modern weapons powered by the "Devil's Poudre"—gunpowder. Those once-invincible walls now presented a target of easy opportunity to the Turks.

Going into the siege, Byzantium knew that it would need modern technology to check its enemies. Considering his situation, Constantine decided to do something to close the gap in the race for medieval weapons of mass destruction. In 1451 he hired a Christian engineer—a Hungarian named Urban—to develop artillery for the Byzantines. But in a poorly considered decision, Constantine, rich as he was, refused to pay Urban adequately. Angered, the cannon expert sallied across the frontier to offer his technological skills to the Turkish side. Mehmet knew a bargain when he saw one and agreed to fully fund Urban's work. Thus, the long tradition of mighty Turkish artillery was born.

Urban's defection had a long-term effect. For hundreds of years—in fact, through the siege of Gallipoli in World War I—Turkish artillery would be among the most advanced and effective in the world.

The formidable Turkish cannon

Once Urban was in the employ of the sultan, he cast the largest cannon ever produced. The mammoth iron tube had a 29-foot-long barrel that fired stones said to weigh around 1,200 pounds. The cannon, as befit a device so large, was given its own name, or even two: it was sometimes called the Imperial and sometimes the Basilica. It was so heavy that a team of 60 oxen had to haul it to Constantinople from its foundry in Edirne, a city 100 miles to the west.

Each time the cannon was fired, the expanding gases and friction wore and stressed its barrel to such an extent that it could be fired only seven times a day.

Nicolo Barbaro was an Italian physician in service to Constantine. An observant diarist and reporter, he kept careful notes about what was happening during the Ottoman siege. Urban's Imperial, he wrote, was very large: "The cannon was of exceptional size, throwing a ball 1,200 pounds in weight, and when it fired the explosion made all the walls of the city shake, and all the ground inside, and even the ships in the harbor felt the vibrations of it.

"Because of the great noise, many women fainted with the shock which the firing of it gave them. No greater cannon than this one was ever seen in the whole pagan world, and it was this that broke down such a great deal of the city walls."

In addition to the cannon, the Turks employed many other techniques associated with medieval siege warfare. They dug tunnels under the walls and built tall siege towers that they rolled up to the walls.

The Byzantines mounted a vigorous defense. They dug counter tunnels and threw burning material into those of the invaders. They poured hot pitch from the walls and set fire to anything wooden set against them. The smoke of the fires, as well as the cannon, meant that the combatants fought without clearly seeing what was around them.

Mehmet had a force of 80,000 men at his disposal, regular troops and reserves, and 12,000 elite forces, perhaps the best fighters in all of Europe and the Middle East. These special fighters were called janissaries. Taken from Christian families as youths, janissaries were converted to Islam, then raised and trained in special schools to become fanatical fighters. They were the sultan's personal guard and most skilled soldiers.

Facing this army, Constantine could muster only about a quarter as many troops. Still, he did have those walls for protection. Perhaps, he thought, he could obtain help from European powers such as Venice, which might be sympathetic in the fight between a Christian empire and an Islamic one. But the Venetians offered only words of encouragement. It became clear to Constantine that Byzantium would have to go it alone.

Sultan Mehmet stopped the cannon bombardment temporarily while an emissary was sent into the city to discuss surrender terms. If the emperor paid an annual tribute of 100,000 gold bezants, the emissary said, then Constantinople would remain Byzantine. Or all the citizens could simply take their possessions and leave, and no one would be harmed.

Unable to raise the money and unwilling to leave the city, Emperor Constantine and the Byzantines decided to tough it out. If Byzantium and Constantinople were to die, they would die fighting.

On May 29, 1453, Mehmet and his soldiers made a final, terrific assault at the location called the San Romano Gate. The battle began with the Turks "firing their cannon again and again, with so many other guns and arrows without number and shouting from these pagans, that the very air seemed to be split apart. . . . The janissaries," Barbaro continued, "and their officers attacked the walls of the poor city, not like Turks but like lions, with such shouting and sounding of castanets that it seemed a thing not of this world. The shouting was heard as far away as Anatolia, twelve miles away from their camp. . . . The people cried at the top of their voices, 'Mercy! Mercy! God send help from Heaven to this Empire of Constantine.' "

Bitterly, Constantine saw that no help was going to come. He had given up hope of assistance coming from the pope, from the Venetians, and now from God. The Turks fired their great cannon again and again. The janissaries poured through the breach in the walls, according to the diarist, "like hounds and came on behind the smoke of the cannon, raging and pressing on each other like wild beasts." Within 15 minutes more than 30,000 Turkish troops were inside the walls, and the city was taken.

Constantine fought and died at the walls of his city. His head, according to several reports, was cut from his body and presented to Mehmet as a trophy. After 1,100 years of power, the Queen of Cities fell to the guns of Urban and the troops of Mehmet the Conqueror. The end had come for Byzantium at the end of a long cannon barrel.

Long-barreled cannons, even those as large and powerful as Mehmet's Imperial, are simple and straightforward devices. Pared down to the basics, a cannon projects a slug or a shell toward a target via a long, straight aiming barrel. Military cannons use gunpowder, cordite, or other chemical propellants to impart energy to the projectile.

In the 15th century Sultan Mehmet and other powerful leaders were certainly taken with their armies' ability to produce the thunderous roar, fire, and smoke associated with gunpowder artillery. Their possession of powerful cannons, and the power that such devices bestowed upon their owners, explain their popularity.

Early gunpowder-powered cannons were expensive, labor intensive, and unwieldy to operate, but they played a key role in the armory of nearly every military power from the mid-14th century onward.

But other types of cannons, ones that use no gunpowder at all, have been tested and used. In fact, there are numerous historical accounts of noncombustion-powered cannons. For example, the energy transferred from a rapidly unwinding spring, a discharging electronic capacitor, or an expanding quantity of steam could also shoot a projectile down a muzzle quite effectively. (In chapter 7 we'll explore the steam cannon in depth.)

In the following project, the energy stored in a flexible rubber membrane is transferred to a pellet to make a small but interesting toy cannon.

The Elastic Zip Cannon

The Elastic Zip Cannon is a cleverly designed toy hand cannon that uses stretched rubber fabric (a balloon) to power a small projectile, such as a dried bean, through a plastic pipe toward a target. It's a simple yet amusing project that consistently delivers good results.

Safety

The Elastic Zip Cannon is a modified slingshot powered by a stretched plastic membrane. The mechanism, while not particularly dangerous, does shoot peas and dried beans with enough velocity to sting. So protect eyes, and don't shoot it at unsuspecting or unwilling participants. Be sure to follow the directions on pages 11–12 for working with PVC solvents and cleaners.

TOOLS

- Hacksaw
- Small-diameter round (rat-tail) hand file
- Medium grit sandpaper

MATERIALS

- (1) 12-inch-long, ½-inch-diameter PVC pipe
- (1) 4-inch-long, ½-inch-diameter PVC pipe
- (1) ¼-inch-long, ¾-inch-diameter PVC pipe
- (1) Can PVC primer
- (1) Can PVC cement

- (1) ½-inch-diameter PVC tee fitting
- (1) ½-inch-diameter PVC end cap
- 12-inch high-quality balloons
- A handful of dried peas or beans. Larger dried beans, such as Limas or garbanzos, work the best.

3.1

BUILD THE ELASTIC ZIP CANNON

The Elastic Zip Cannon is quite straightforward and easy to make.

1. Use the hacksaw to cut a 12-inch and a 4-inch length of ½-inch PVC pipe.
2. Using the file, bevel the inner edges of the opening on one end of the 12-inch piece. A long, smooth transition works best. **See diagram 3.1 inset.**
3. Take the ¼-inch length of ¾-inch-diameter PVC pipe and slide it like a ring onto the 12-inch piece. You may need to slightly enlarge the diameter of the ¾-inch-diameter ring with sandpaper if necessary in order to slide it onto the 12-inch piece.
4. Position the ¾-inch-diameter ring about ½ inch from the beveled end, and solvent-weld it into place using PVC primer and cement (see page 11–12).
5. Attach the PVC tee fitting to the other end of the 12-inch-long piece as shown in **diagram 3.1**. Attach the 4-inch-long PVC handle to the tee. Then attach the end cap to the other end of the handle. Solvent-weld each piece into place.

3.2

6. Pull the mouth of the balloon over the ¾-inch diameter ring. Align the balloon such that its center line aligns with the center line of the 12-inch-long PVC barrel. The friction and tight fit will hold the balloon onto the barrel.

OPERATING THE ELASTIC ZIP CANNON

3.3

1. Hold the cannon vertically. Insert a bean into the barrel, and let it fall down into the balloon.
2. Level your shooter. Pinch the bean inside the end of the balloon, and pull the balloon back so it stretches.
3. Aim the cannon at a target and release. The bean will shoot toward the target.

PROBLEMS?

If the bean gets stuck inside the barrel:

- If the inside edge of the barrel, underneath the balloon, has any sharp edge at all, the bean will strike it and break apart or get stuck. Make sure the leading edge is smooth.

If the balloon breaks:

- Balloons don't last forever and will break after repeated use. Balloons are cheap, so just replace them as required. But if your balloons break immediately, you may be pulling too hard or you may be using poor-quality balloons.

SPORTS SCIENCE AND THE MECHANICAL TOE

Perfection is a rare thing, and examples of perfection in sports in particular are few and far between. But some actions are truly perfect because they are as good as they can possibly be. Gymnast Nadia Comaneci received the first perfect score awarded in an Olympic competition in 1976. In baseball, a perfect game means the pitcher handled every single batter he faced without allowing a hit, a walk, or an error. Literally, no one can do it better. Hundreds of thousands of major league baseball games have been played since 1880, and in that time there have been only a couple dozen perfect games. Very few bowlers, even those who bowl frequently, have rolled a 300 game. Those who have done so have attained perfection; no one will ever score better.

Football doesn't lend itself to perfection. There's always a chance that someone will score more points or do something better, faster, longer. Except for one record.

Back in 1969 the world champion New York Jets, led by Joe Namath, took on the Broncos at Mile High Stadium in Denver. The heavily favored Jets spent the first half of the game manhandling the

Broncos. They built up a 15-point lead before the Broncos roared back and beat the Jets 21–19. The stunning loss was the talk of the sports world in the next morning's papers.

Almost lost in the big stories was the fact that little-known Jets punter Steve O'Neal set a record that still stands, one that may be as hard for future athletes to break as Joe DiMaggio's consecutive-game hitting streak, or Wilt Chamberlain's 100 points in a single basketball game. O'Neal, a Jets rookie, was sent in to punt the ball to the Broncos. With the ball on his own one-yard line, he was in a tough position. Normally an NFL punter stands 15 yards behind the center. But with the ball on the one, he could stand back only 11 yards from the onrushing linemen.

The air is thin in Denver, and O'Neal—a good but in the final analysis not great punter—really got his foot into it. The ball traveled in an arching spiral trajectory for nearly 70 yards, sailing over punt returner Bill Thompson's head. It bounced near the Bronco 30-yard line and finally rolled to a stop. On the one-yard line. O'Neal set a pro football record with a 98-yard punt that was 4 yards better than the previous record set in 1923 by Wilbur "Fat" Henry with the Canton Ohio Bulldogs, then in the National Football League.

If you think about it, 98 yards is a perfect punt. It is as far as a ball can be punted. Since half-yards are not recorded in football records, a punt of 99 yards or more from the 1-yard line would result in a touchback, netting only 79 yards.

The Mechanical Toe

The Mechanical Toe is a simple and oddly amusing device for propelling miniature footballs and tennis balls around the backyard. Unquestionably, you can throw a football farther than this contraption can toss it, but that's not really the point. The Mechanical Toe is easy to build, fun to use, and lends itself to a variety of experiments, some scientific and some simply amusing.

Safety

To kick a miniature football, the Mechanical Toe simulates the action of the human leg. The Mechanical Toe's power source is a stretched elastic cord, usually called a bungee cord. Bungee cords are capable of storing considerable energy when stretched. You must make certain that your cord is securely attached and in good condition prior to using the Mechanical Toe.

1. Wear safety glasses and gloves, at least during the initial test and the first few trials.
2. Clear the area in front of the Mechanical Toe, especially during test kicks, as the toe piece could come loose if not properly cemented.
3. Be sure to keep hands and feet clear from the impact area under the handle at all times.

TOOLS

- Handsaw
- Plastic spoon
- Hammer
- Power drill
- 11/32-inch-diameter drill bit
- Phillips head bit (if deck screws are used)
- (1) 1-inch-diameter wood boring bit
- Stapler or glue

MATERIALS

- (3) 6-inch-long, 1-inch-diameter PVC pipes
- (2) 2-inch-long, 1-inch-diameter PVC pipes
- (1) 1-inch-diameter PVC tee fitting
- (3) 1-inch-diameter PVC socket style (no pipe thread) end caps
- (1) 1-lb. box plaster of Paris
- (1) Can PVC primer
- (1) Can PVC cement
- (1) 36-inch-long wood 2 x 4 (the base)

- (2) 12-inch-long wood 2 x 4s (the uprights)
- (1) 12-inch-long wood 2 x 4
- (2) 6-inch-long wood 2 x 4s
- 1-lb. box 2½-inch nails or deck screws
- (2) size #2 eye-screws
- (1) 2-inch by 2-inch thick foam or felt pad
- (1) 90-degree 1-inch-diameter PVC elbow fitting
- (2) 45-degree 1-inch-diameter PVC elbow fittings
- (2) 2½-inch-long #8 machine screws with 4 nuts
- (1) 7-inch-long 5⁄16-inch-diameter, fully threaded bolt, 4 washers, and 3 nuts
- (2) 18-inch bungee cords, loops or hooks at each end
- Assorted lengths of PVC pipe for holding the football or other balls

BUILD THE MECHANICAL TOE

4.1

1. Cut PVC pieces to length. Measure and mark cutting lines for the three 6-inch PVC pipes and two 2-inch PVC pipes. Use the saw to cleanly and squarely cut the pipe.
2. Make the kicking toe by solvent-welding one 2-inch-long pipe into each opposing end of the tee. (Leave the center hole of the tee open for now.) Solvent-weld an end cap onto each pipe end. (See pages 11–12 for solvent-welding instructions.)

3. Following the directions on the package, mix a sufficient quantity of plaster of Paris to fill the tee and end cap assembly. Using a plastic spoon, fill the tee–end cap assembly with the plaster of Paris.

4.2 Kicking Toe Assembly

4. Before the plaster of Paris hardens, solvent-weld a 6-inch pipe into the middle hole of the tee. Put the toe aside, and allow it to dry and harden.

BUILD THE WOODEN BASE

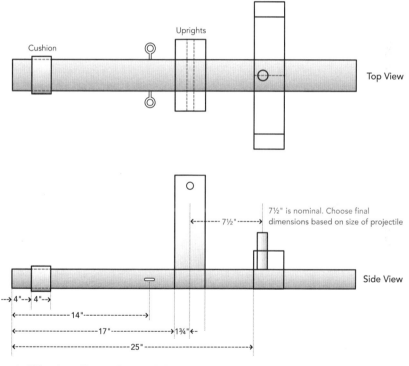

4.3 Wooden Base Assembly

1. Next, build the wooden base. Cut the 2 x 4s into lengths as described in the material list.

2. Attach the uprights to the base, using nails or deck screws. See **diagram 4.3**.

3. Drill two $^{11}/_{32}$-inch holes through the uprights.

4. Carefully note the location of the upper stabilizer on the base as shown on **diagram 4.3**. Attach the upper stabilizers to the base using nails or deck screws.

4.4

5. Slide the lower stabilizers into position, and attach them to the upper stabilizer with nails or deck screws.

6. Attach the eye-screws to the base at the location specified in **diagram 4.3**. Affix the foam or felt pad cushion to the base with staples or glue at the location shown in **diagram 4.8**.

BUILD THE KICKING LEG

1. Drill one clean $^{11}/_{32}$-inch hole for the pivot bolt through the 90-degree elbow fitting.

2. Solvent-weld the leg assembly together as shown in **diagram 4.6** using the assembled toe

4.5

assembly, the 90-degree elbow fitting, the two 45-degree elbow fittings, two 6-inch PVC pipes, and one remaining PVC end cap. Be certain to fully seat all connections and align all joints with the plane formed by the uprights and braces.

4.6 Kicking Leg Assembly

3. As shown in **diagram 4.6**, drill two ³⁄₁₆-inch-diameter holes through the leg assembly for the 2½-inch-long, #8 machine screws. The two holes are perpendicular to the ¹¹⁄₃₂-inch pivot bolt hole.

4. Insert the two #8 machine screws through the holes and bolt them into place. The space between the protruding machine screw heads is designed to keep the bungee cord positioned at the correct spot on the leg assembly.

ASSEMBLE THE MECHANICAL TOE

1. Align the holes in the base with the hole in the leg assembly. Insert the ⁵⁄₁₆-inch-diameter bolt. Place the washers and nuts as shown in **diagram 4.7**, and tighten the nuts.

2. Drill a 1-inch hole about ½ inch deep in

Bolt Head
Washer
Nut
Washer
PVC Elbow
Washer
Nut
Washer
Nut

4.7

4.8

the middle of the upper stabilizer, at the point where the fully down toe will almost, but not quite, strike it.

   Note: It may take a bit of trial and error to find the best point to drill the hole for the kicking tee. Place the assembled kicking leg in position on the wooden base prior to drilling the hole for the tee, and make certain the toe end will strike the projectile in the desired spot.

3. Place the 3-inch-long, 1-inch-diameter pipe in the hole. Wrap the bottom half-inch of the pipe with duct tape if necessary to obtain a good fit in the hole. This is the kicking tee. The tee will hold the ball or other projectile. PVC tees of other lengths will be needed for different projectile sizes. This arrangement will allow you to quickly and easily swap tees to accommodate different objects.

4. Attach the bungee cord. Place the center of the bungee cord between the protruding screw heads on the leg assembly. Attach the hooks on the bungee cord to each eye-bolt on the base.

4.9

5. Test the alignment. Be certain the handle hits the cushioned area on the base. Check to be certain the bungee cord stays between the cord holder made from the #8 machine screws at all times.
6. Make sure the arm assembly pivots freely, but without excessive wobble or travel on the pivot bolt. Make centering and tightening adjustments as necessary.

OPERATING THE MECHANICAL TOE

1. Place a football, handball, tennis ball, or other object on a kicking tee.
2. Lift the arm. As you do so, the bungee cord will stretch. Release the arm to let the toe strike and launch the projectile.

MODIFICATIONS

To increase the Mechanical Toe's performance, try several modifications:

1. Use a thicker bungee cord, or use two medium-thickness bungees.
2. Increase the weight of the toe.

3. Increase the radial velocity of the toe by lengthening the leg. Doing so will change the geometry of the toe. This will require taller uprights and a longer bungee.
4. Use a harder ball.
5. Place the ball slightly off center to impart a spin to it. A spinning ball will travel farther.

ISAAC NEWTON AND FOOTBALL

Analyze any type of sporting activity, and sooner or later you will find yourself knocking up against some pretty basic physics. After all, a sport is often defined as "an activity that involves physical exertion and physical skill." Notice that you can hardly write *physical* without first writing *physics*.

The Mechanical Toe is a way to study the very human act of kicking a football using a mechanical model. The Toe lays bare the physics of kicking by doing away with the cumbersome human biology of skin, muscles, joints, foot bones, knee bones, and ankle bones. It allows you to focus explicitly on the physics.

Once a football leaves the foot or the hand of a player, two forces act on it: gravity and the drag caused by air rushing over its surface. In general, this rule holds true for all thrown, kicked, batted, clubbed, or otherwise flying balls. It is true in particular for football's basic motive modalities: the kick, punt, and pass.

What accounts for the different trajectories of kicks, punts, and passes? Many factors influence a football's flight, specifically, how far it travels, the time it spends in the air (in football parlance, its hang time), and how accurately it is directed to the spot where the punter desires it to go. Those factors are:

- Launch speed
- Launch angle
- Air density and wind
- Spin of the football
- Shape of the football
- Surface of the football

In the 17th century Isaac Newton first described the laws of nature that account for the behavior of objects in motion. These fundamental physics concepts are now known as Newton's three laws of motion. These simple principles underlie all aspects of classical physics dealing with the motion of any object larger than an atom and going slower than near–light speed. Newton's third law of motion is:

> In every interaction, there is a pair of forces acting on the two interacting objects. The size of the force on the first object equals the size of the force on the second object. The direction of the force on the first object is opposite to the direction of the force on the second object.

The collision of the Mechanical Toe with the football is an interaction between two objects that have made intimate and robust contact with each other. So how does Newton's third law of motion apply here? In the collision between the Mechanical Toe and the ball, both objects experience forces that are equal in magnitude and opposite in direction. The forces cause one object to speed up and the other object to slow down.

While the forces are equal in magnitude and opposite in direction, the acceleration of the objects is not equal. Because the toe and the ball have different masses, their collision results in unequal accelerations. The lighter object will have the larger acceleration.

If we could fill the Mechanical Toe with lead instead of plaster, then the more massive toe would impart even greater acceleration to the ball. If we could fill the toe with the same volume of an ultraheavy element such as thorium or uranium (plutonium would be even better!), then the toe would have even more mass, and the ball would really go flying!

Now let's analyze this kick by looking at Newton's other laws. Newton's first law says:

> An object in motion will stay in motion and an object at rest will stay at rest unless acted on by a force. That is, once an object is in motion, it stays in motion until something happens to change it.

Newton defines a property of objects in motion called momentum, which is a consequence of its combined mass and velocity. The bigger and faster something is, the more momentum it has. So it is not incorrect to say that the game of football is mostly about using and controlling momentum, and coaches know this.

On most football teams, the punter is not an especially big guy—he'll be around 190 pounds, maybe 200. That may seem like a lot until you consider the players weighing 300 pounds who want to rush in and pound him to the ground before he can get his kick away. As soon as the center hikes the football, a stampede of defensive linemen make for the punter with a single aim: to slam him to the ground as quickly as possible.

Of course, there is an obstacle in the way of those defensive players. They have to push past the punter's cordon of beefy bodyguards—big, brawny guards, tackles, and ends who weigh an average 20 pounds more than even the defensive linemen.

To navigate through games with their incredible mass, skillful offensive linemen know a lot about leverage, physics, and momentum, although they may not think about these things in those terms. The players with the most mass have a weighty advantage. Their big muscles and even bigger gut help them keep the defensive players away from the punter. A really big football player can weigh more than 350 pounds, and he uses his mass like a physics pro to do his job, keeping his center of gravity low, his feet wide apart, and his shoulders in front of his body. Imagine attempting to push aside a moving 350-pounder in full pads. That's hard to do, because he's got a lot of momentum!

Newton's second law of motion is just as important as the other two when it comes to football, for this is the law that describes the relationship between force and mass:

$$\text{Force (F)} = \text{mass (m)} \times \text{acceleration (a)}$$

Acceleration is the change in speed or direction as a result of force. The second law says bigger football players are harder to move than less massive football players. The third law of motion says that the bigger the lineman, the more force is required to change that offensive lineman's speed or direction.

You may want to think of two grocery carts sitting in the super-market aisle, one empty and one filled with food. The cart stuffed with groceries has more mass and will therefore be harder to move. Moving it takes more force. Likewise, making those giant football tackles change direction or speed takes more force. So a less massive player will have a tough time pushing aside a big offensive tackle to get his hands on the punter.

LEGENDARY LAUNCHERS

Suppose an experimenter has a small metal tube with a tight-fitting cover for each end. On a whim, she takes the metal tube and puts some fuel and oxidizer inside it. She tightly covers and fastens both ends and then ignites the fuel by applying heat or lighting a fuse. When enough energy is provided, or the burning fuse contacts the fuel-oxidizer mixture, the chemicals combine in a vigorous exothermic, or energy-producing, reaction. Rapidly expanding gas is created, and a great and perhaps uncontainable pressure builds up inside the tube. If the pressure exceeds the material strength of the tube, the whole thing blows up. This is, simply put, a pipe bomb.

Now instead of solidly fastening covers to both ends of the tube, suppose the daring experimenter simply slides a movable cover over one end. When she ignites the fuel, now what happens? Most likely the sliding cover flies off. It's a projectile, and she has made a primitive cannon.

The experimenter next leaves one end of the tube completely open and ignites the fuel while the tube lies upon the ground. When the chemicals ignite, the burning gas rushes out of the open end and propels the tube in the other direction. She has just built a rocket.

A cannon, a bomb, and a rocket are merely different manifestations of the same basic idea. The form that the device takes is simply a matter of details.

THE AIR BULLET

In an upcoming chapter, we will see how pressurized air can be used to project items through the air for considerable distances. For now let's consider another type of cannon, one in which air itself is the projectile.

It's hard to believe, but air (or smoke, or another vapor) itself can be made into a bullet of sorts. We're not imagining a simple stream of jetted air, but rather a real tangible projectile: an air bullet. In this chapter we will create a device called a Vortex Launcher that shoots a projectile made of air. The air bullet can be made visible, and you'll see that the bullet has a discrete size and shape and that it shoots through the air like an arrow (albeit a slow-moving arrow shaped like a doughnut).

The Vortex Launcher is capable of shooting a ring of air or smoke from one side of a room to another. It can blow out candles, knock down a house of cards, and set curtains aflutter. That's mere puffery compared to what the US Army once conceptualized: an air bullet and vortex projector big enough to knock airplanes out of the sky.

During World War II the US Army was interested in developing alternative methods of dispersing gases and vapors on the battlefield. The ugly truth is that the purpose of these experiments was to figure out "better" ways to use chemical weapons, although this technology might also have useful agricultural applications such as spreading insecticides. One method the army studied involved forming gases and vapors into high-speed traveling rings called vortexes. A vortex has the useful property of maintaining its shape—a doughnut—while it flies through the air.

Early in the war the army hired a well-known aircraft designer named Thomas Shelton to build a prototype vortex generator to prove the concept. Shelton was a highly regarded aeronautical engineer, well known for his futuristic wing and fuselage designs. In 1935 Shelton had become famous in aerodynamic engineering circles for designing the Crusader, a small, low-powered, teardrop-shaped airplane with two tails that was capable of cruising at 200 miles per hour.

For the US Army project, Shelton built a bell-shaped chamber with a piston powered by a shotgun shell on one end and a hole in the other. He fired off the shotgun shell inside the chamber, and some pretty interesting things happened. An eighteen-inch-diameter ring of smoke shot through the air—traveling a full city block and making a weird howling sound as it whizzed by.

Shelton proposed building a larger and more powerful vortex launcher for use as an antiaircraft weapon. He suggested building one with a piston powered by a high-explosive charge. According to Shelton, such a vortex weapon could accurately fire a ring of air at supersonic speeds to an altitude of several thousand feet. The vortex would hit enemy aircraft and send them spinning, in theory causing them to lose control and knocking them out of the sky.

For a while the army funded some research into the idea. But ultimately it decided its research money should be spent elsewhere, and an air bullet capable of taking down an airplane was never developed. Nonetheless, the vortex gun did show up on a smaller scale in several toys, including the Flash Gordon Ray Gun (named toy of the year in 1949 by *Popular Mechanics*) and the Wham-O Air Blaster.

The Vortex Launcher

This project is simple and inexpensive, but it shoots a pretty darn good vortex. And the bigger the vortex launcher, the bigger the vortex.

TOOLS

- Large scissors
- Electric drill
- 3-inch-diameter hole saw*
- ³⁄₁₆-inch-diameter drill

---

* The exact size of the hole is not critical, but having a perfectly circular hole is important. A 3-inch hole saw is commonly available at hardware stores for cutting holes in doors for doorknobs.

MATERIALS

- 14-inch-diameter round piece of thick plastic or rubber sheeting
- Large plastic bucket, roughly 12 inches in diameter and 15 inches long (these dimensions are not critical)
- 3-inch long, ¼-inch-diameter eye bolt with two nuts and two washers
- Duct tape
- Optional targets and launching materials (see pages 77–78)

MAKING THE VORTEX LAUNCHER

5.1 Vortex Launcher Assembly

1. From the piece of thick plastic or rubber sheeting, cut a circle of a diameter approximately 2 inches larger than the diameter of the open end of the plastic bucket.
2. Drill a 3-inch hole in the center of the bottom of the bucket.
3. At 2-inch intervals, use the scissors to cut radial lines into the plastic or rubber sheeting extending ¾ inch toward the center.
4. Drill a ³⁄₁₆-inch hole in the center of the sheeting. Thread one of the nuts and one of the washers onto the eye bolt. Push the bolt through the opening, and place the other washer and nut on the bolt.

5. With the eye of the
   bolt on the outside,
   duct-tape the flexible
   sheet to the opening.
   Make certain that
   the perimeter is well
   sealed. The sheet
   should be attached
   such that its center can
   be pulled back several
   inches. But don't make
   the sheet too loose or it won't move air efficiently. You may need
   to experiment a bit to find the right balance.

5.2

OPERATING THE VORTEX LAUNCHER

1. To operate the Vortex Launcher, pull back on the eye bolt, and
   then quickly push it forward. How quickly should you operate the
   sheeting membrane? How far back should you pull, and how far
   forward should you push? The optimal technique is best learned
   by trial and error. Take note of the method that produces the
   tightest vortexes and best results.

INTERESTING OPTIONS

Target Practice

For practice, set up small houses of playing cards, and figure out the
best way to knock them down.

Visible Vortexes

The easiest way to see the vortexes as they shoot by is to fill the
interior of the bucket with smoke. Blow smoke from a cigar into the
bucket and then launch the vortex. I spray a fine mist of water into
the interior along with the smoke. The moisture seems to make the
rings hold together longer.

For nonsmokers, dry ice works well. Submerge a small hunk of dry ice in a container of water, and insert it into the cannon. This will produce copious visible condensation. Dry ice is dangerously cold and injures unprotected skin; always handle it cautiously and wear gloves.

Odoriferous Projectiles

Because the launcher has pretty good accuracy and launches fairly tight vortexes, you can direct a projectile of smell at unsuspecting individuals using the Vortex Launcher.

Place cotton that has been soaked with essential oils, such as peppermint or lavender oil (often available at drugstores, cosmetics stores, and natural foods co-ops). Let the smell concentrate for a minute inside the Vortex Launcher, then shoot.

THE HELLBURNER OF ANTWERP

The lesson that Emperor Constantine XI learned the hard way (see chapter 3)—that good engineers deserve good pay, especially if they are able to build machines that threaten your empire—is something that leaders should always remember in times of war. But the lessons of history are often forgotten. And evidently this same lesson was lost on the Spanish king Philip II.

Federico Gianibelli of Mantua, Italy, was a military engineer of exceeding skill and cleverness. Around 1580 he was rudely rebuffed by Philip and the Spanish royal court when he offered them his engineering services. Angered by the patronizing attitude of the Spanish king, he went to Antwerp to work for the Dutch, who far more politely asked him to devise a defense to protect the city from an attack by the Spanish Armada. They backed up their request with cold, hard cash.

Gianibelli's device was the terrible-beyond-description "Hellburner." The Hellburner was a huge floating time bomb. A shell of a ship packed to the gills with explosives, incendiaries, and shrapnel, the Hellburner was designed to float into the middle of the besieging Spanish fleet with a time-delay fuse.

John Lothrop Motley, a 19th-century British historian, describes Gianibelli as "purely a man of science and of great acquirements, who was looked upon by the ignorant populace alternately as a dreamer and a wizard. He was as indifferent to the cause of freedom as of despotism, but he had a great love for chemistry. He was also a profound mechanician, second to no man of his age in theoretic and practical engineering."

Gianibelli's plan for defending Antwerp involved two ships, weighing approximately 75 tons apiece, which he requisitioned from the Dutch sailing fleet. The two ships were renamed *Fortune* and *Hope,* ironic names considering their destructive purpose. They were to become wave-borne volcanoes.

Gianibelli ordered workmen to place a solid 12-inch-thick floor in the hold of each vessel. Built up the ship's sides from the floor was a bowl of masonry 40 feet long and 3 feet high. The men working on the ship called this bowl "the crater."

The crater was filled with several thousand pounds of explosives, but not just any explosives. They used a new and secret type of gunpowder that had been compounded and ground by Gianibelli himself. Over the deadly crater workmen built a marble covering in the shape of a hollow pyramid. The pyramid was filled with stones, cannonballs, harpoons, iron hooks, random hunks of jagged metal, and every other available piece of junk that could be used as shrapnel. A wooden roof was constructed to camouflage the devilish payload.

Gianibelli's plan was fairly straightforward. The two Hellburners were to be steered down Antwerp harbor toward the massed Spanish Armada. At the right moment the explosives in the holds were to be detonated by an ingenious timing device, also designed by Gianibelli. The timer would cause a large spark from a flint and steel mechanism to ignite the mass of gunpowder in the ships' holds.

The weather on that momentous day in 1585 was mild. The sky was clear when the Dutch released the *Hope* and the *Fortune* from their moorings, sending them floating slowly on the tidal current down the harbor toward the anchored armada. Spanish naval officers saw the ships as they moved closer. Suspecting that they were in fact some type of floating bomb, they put their own ships and troops on alert. The *Fortune* sailed in first but ran aground near a bridge held by the Spanish. The Spanish watched uneasily as a small fire started on the deck of the ship. But it soon extinguished itself, and no further activity was visible.

The Spanish troops, peering out from behind defensive fortifications, started to laugh and jeer at the Dutch army's poor effort. But they also watched, perhaps a bit nervously, as the second ship, the *Hope,* floated toward them. The *Hope* lazily and innocuously rode the current toward the Spanish position, striking the bridge directly and coming to a stop.

Historians tell us that some of the Spanish sailors saw a thin curl of dark smoke rising from a fire burning on the *Hope's* deck. Officers sent a team in small boats to investigate.

When the team boarded the ship, the mechanical clock controlling the fuse timed out, and a tremendous fireball erupted over the harbor. The *Hope* vaporized, as did the Spanish sailors on board, the bridge, and most of the Spanish troops, even those behind the fortifications. Motley describes the scene and its aftermath in detail:

> Houses were toppled down miles away, and not a living thing, even in remote places, could keep its feet. The air was filled with a rain of plough-shares, gravestones, and marble balls, intermixed with the heads, limbs, and bodies, of what had been human beings. Slabs of granite, vomited by the flaming ship, were found afterwards at a league's distance, and buried deep in the earth. A thousand soldiers were destroyed in a second of time; many of them being torn to shreds, beyond even the semblance of humanity.
>
> The *Hope* had then destroyed herself, precisely at the right moment. All the effects, and more than all, that had been predicted by the Mantuan wizard [Gianibelli] had come to pass. The famous bridge was cleft through and through, and a thousand picked men were blown out of existence. The

Governor-General himself was lying stark and stiff upon the bridge which he said should be his triumphal monument or his tomb. His most distinguished officers were dead, and all the survivors were dumb and blind with astonishment at the unheard of convulsion.

The passage was open for the fleet, and the fleet [of Dutch allies], lay below with sails spread, and oars in the rowlocks, only waiting for the signal to bear up at once to the scene of action, to smite out of existence all that remained of the splendid structure, and to carry relief and triumph into Antwerp.

The Clothespin Snap Shooter

Gianibelli's contribution to military technology was a result less of innovation and invention than of his ability to make use of existing technologies in dramatic and powerful new ways. Fire ships had been used previously, but Gianibelli extended their range through clever innovations in everything from complex fusing to timers to the composition of the explosive.

Similarly, people have been making toy guns out of clothespins probably since spring-operated clothespins were first invented. Most often the pincer-like action of the clothespin is used as a trigger to release a stretched rubber band. The results are adequate but not terribly impressive. But like Gianibelli's Hellburners, the Clothespin Snap Shooter marks a definite upgrade in performance and sophistication over previous clothespin guns. The Snap Shooter makes several novel enhancements to a well-known device that turns it into a better-performing version.

Just as Gianibelli figured out how to go from boat to bomb, this chapter will show you how to go from clothespin to cannon. This project costs only pennies and takes less than a half-hour to build. Be forewarned, however, that it is almost addictively amusing. Upon seeing it work, most people will want to build their own immediately, so have a good supply of clothespins on hand.

The Snap Shooter has been around for a long time, perhaps not when the Spanish Armada ruled the waves, but certainly in the early 1970s. That's when Gordon Wells and Jerome Pohlen, from whom these plans originally came, built it as elementary school experimenters.

It's amazing but true: the lowly spring-operated clothespin provides the perfect raw material for a gun that shoots small projectiles very well. A couple of quick part inversions, a small notch with a jackknife, and a long piece of tape are all it takes.

The Clothespin Snap Shooter is versatile. It shoots wooden toothpicks, Popsicle sticks, dried beans, or just about anything else that fits between the jaws of the gun and is shaped so that it can be flung out by the force of the rapidly uncoiling torsion spring.

Safety

1. The Clothespin Snap Shooter easily shoots across a fair-sized room. Don't aim it at people or breakable items.
2. The spring can pinch fingers when fired.

TOOLS

- Utility knife
- Scissors

MATERIALS

- (2) Wooden clothespins (the type with a coil spring that pinches two separate wooden levers together)
- (1) 3-inch-by-5-inch index card
- Clear packing tape

MAKING THE CLOTHESPIN SNAP SHOOTER

1. Take apart the two clothespins, and lay the parts down on a table. There will be four wooden levers and two springs. Discard one of the springs.

5.3

2. On one of the clothespins, rework the shape of the semicircular notch with a utility knife until the notch has a vertical front end as shown in **diagram 5.5**. When the notch is shaped correctly, the end of the coil spring should hook just inside and stay without slipping.

3. With the scissors, cut two or three small rectangles of cardboard from the index card to form a spacer. Each rectangle should be the width of the clothespin, and the length of the distance from the thin end of the clothespin to the start of the semicircular notch.

5.4

4. Slide the spring onto the modified wooden piece, as shown in **diagram 5.4**. One of the spring's bent ends will rest in the squared-off groove you just made in the wooden clothespin piece in step 2 when the shooter is cocked.

Paper Spacer

Vertical End

Tape

5.5

5. Arrange the two clothespin pieces and the cardboard as shown in **diagram 5.5**. Tape the ends into place using a few turns of strong packaging tape. That's it! The Clothespin Snap Shooter is ready for action.

FIRING THE SNAP SHOOTER

5.6

1. Cock the spring by taking the remaining wooden clothespin piece and pushing the bent metal end of the spring toward the semicircular notches. When the bent end of the spring reaches the notch, it will enter the notch and latch into place. The device is armed and ready to fire.

  Troubleshooting tip: If the bent spring end will not stay in the semicircular groove, add more layers of cardboard between the two flat clothespin pieces. Doing so will change the geometry of the notch. You may need to experiment with the cardboard and the angle of the notch until the geometry is just right.

2. Trigger the shooter by gently pulling on the spring coil. When the spring leaves the notch, it pushes against the projectile, firing it across the room.

5.7

STADIUM PSYCHOLOGY AND THE T-SHIRT CANNON

It is not always easy to determine the true provenance of a world-changing invention. In fact, the history of invention and discovery is frequently clouded by controversy as to which person or group deserves credit. For instance, Alexander Graham Bell is widely believed to be the inventor of the telephone, but a case can be made that it was developed earlier by others. There are quite a few claimants, including Elisha Gray, Johann Philipp Reis, and Antonio Meucci.

Another inventive controversy concerns the television. Two individuals claim paternity: Vladimir Zworykin, a Russian-born American inventor working for Westinghouse, and Philo Farnsworth, a privately backed inventor from Utah. Yet another controversy, well known among historians, involves the invention of radio. While most history teachers give credit to Guglielmo Marconi, a very strong case based on patent information can be made for Nikola Tesla, who claimed that he was truly the father of radio.

Such disputes are familiar to those who study the history of science. When it comes to television, radio, and telephones, establishing the rights to such products involves huge amounts of money. But controversy can arise even for the invention of things whose only value is bragging rights. Case in point: the Wave.

The Wave is a well-known phenomenon commonly seen at large, crowded sporting events. An exercise in crowd psychology, it involves a wave of jumping humanity sweeping around a stadium or sports arena. The Wave starts when a group of people leap to their feet with their arms up and then immediately sit down. Under the right circumstances, people next to the initiators do the same thing, as do those next to them, and so on. Viewed from above, a synchronized wavelike movement rolls through the audience, momentarily disturbing the equilibrium like a gust of wind across a field of Dakota wheat.

Two different groups claim to have invented the Wave, and both attempt to back up their claim.

While one frequent claim is that the Wave was first performed in Mexico City during the 1986 World Cup, it was almost certainly in use at sporting events in the United States prior to that.

George Henderson (aka Krazy George), a professional cheerleader, claims he invented the Wave at an Oakland Athletics baseball game in October 1981. He states on his website (www.krazygeorge. com), "I, Krazy George, invented, created and otherwise started the most famous fan participation routine in the history of the world."

Opposing Krazy George is a group of University of Washington alumni who say their school's cheerleaders orchestrated the first Wave during the third quarter of a Washington-Stanford football game, also in October 1981.

Even if determining with certainty the one true date and place where it first began is impossible, the Wave is interesting in ways beyond its ability to exhort teams to play better or to keep fans interested in a lopsided game. It reveals interesting relationships between human psychology and the physics of waves.

Professor Tamás Vicsek of Eötvös Loránd University in Budapest, Hungary, analyzed videotapes of 14 waves in large football stadiums. He found that human waves can be modeled mathematically and conform to rather rigid physical rules of behavior, even though they are composed of the short-lived actions of thousands of people with no personal connection to one another, save the fact they root for a particular sports team. Vicsek found that waves usually propagate in a clockwise direction, move at a speed of about 27 miles per hour, and are about 25 feet wide, which corresponds to an average width of 15 seats.

New research pertains to the critical mass of wave initiation: What does it take to start a wave? Specifically, how many people does it take to start one? The answer lies in a property that Vicsek terms the "activation threshold."

The activation threshold is a measure of how likely the average person in a crowd is to stand up and do the Wave if the people around him or her stand up. The activation threshold is affected by a host of factors including the density of the crowd (i.e., the number of empty seats) and the level of excitement within the stadium.

Vicsek found that when conditions are right, the Wave can be initiated by fewer than two dozen people standing up simultaneously. In less favorable conditions, an initial group of 40 people may be required to start the phenomenon.

Students of the stadium wave phenomenon may ask, Why do stadium waves go clockwise? Does this have something to do with the Coriolis effect? Do stadium waves travel counterclockwise in the southern hemisphere? No, they almost always go clockwise because most people are right-handed.

The T-Shirt Cannon

In addition to the Wave, other nonfootball-related phenomena occur at events that involve cheerleaders. Many observers feel that few things excite crowds at sporting events more than getting something, anything, for free. At the pinnacle of this measure is the moment when a cheerleader marches onto the field with perhaps the most coveted of all human possessions: a free T-shirt.

At one time the delivery of valuable booty was limited to the distance a human arm could throw it. This limitation provided spectators in rows 1 through 15 with an inestimable advantage. But with modern technology, even a fan in the nosebleed section has a chance to grab a hurled T-shirt. The enabling technology? The T-shirt cannon.

Although the word *cannon* sounds dangerous, no statistical records detail people being injured by a ballistic T-shirt. That's not to say, however, that no one has been hurt in stampedes to grab them.*

Presented here are plans for designing and fabricating a T-Shirt Cannon. May your aim be true and your trigger finger quick.

Safety

Many of the same safety and legal concerns associated with the Night Lighter 36 (see chapter 1) hold true for the T-Shirt Cannon. For example, T-Shirt Cannons may not be legal in your area. So again, you should first check with local authorities to see if any regulations pertain to the construction and use of T-Shirt Cannons.

Second, devices that use pressurized air can be dangerous. They can shoot with great power, and in the worst-case scenario they can fail dangerously.

1. The T-Shirt Cannon is designed to shoot T-shirts. Shooting any other objects, especially near people, is not recommended. (However, I have tried shooting foil-wrapped hot dogs with some success.)

2. Do not fire directly at people. If the cannon is used to distribute T-shirts to crowds, use a high firing angle when aiming the gun so the shirt falls down gently on people. Make sure people are aware of your activities before you use the gun. Hitting an unprepared person, even with something as soft as a T-shirt, can cause injury.

3. Because of the high air pressure inside the tank, always use care and wear eye protection.

---

* Maude Flanders, a fictional character on television's *The Simpsons,* was fatally injured in a February 2000 episode in a freak T-shirt cannon incident.

4. Don't operate a damaged or worn T-Shirt Cannon. Check the cannon frequently for signs of wear.

5. Don't take any construction shortcuts, use inferior materials, or make substitutions in the materials list.

6. PVC's mechanical properties become impaired in extreme weather. Don't use a T-Shirt Cannon in temperatures below 55° or above 95° F.

7. The vapors from PVC cement are flammable. Allow all joints to dry fully before using.

8. Wear safety glasses as you build and operate this device

9. In general, despite best efforts, things occasionally go wrong. If you follow the directions here carefully, the chances of a bad outcome are remote, but unexpected things can happen: material flaws, bad judgment, pure bad luck. Always use common sense, and take careful note: *Build and use this device at your own risk.*

Working with PVC Pipe

Rules and techniques for working with PVC pipe were provided on pages 9–12. Review those instructions before undertaking this project.

It is especially important to review the general safety instructions for PVC solvent, which are on **page 11**. Finally, read and follow all chemical instruction labels carefully.

TOOLS

- Tape measure
- Hacksaw
- Power drill
- Drill bit (sized to fit tank valve)
- Screwdriver
- Pliers
- Utility knife
- Wire stripper
- Soldering iron (recommended)
- Safety glasses
- Broomstick or sturdy dowel
- High-pressure air source (e.g., bicycle pump or air compressor)

MATERIALS

The Barrel Assembly

- (1) 25-inch-long, 3-inch-diameter schedule 40 PVC pipe
- (1) 32-inch-long, 2-inch-diameter schedule 40 PVC pipe
- (1) 3-inch-to-1-inch PVC reducing coupling, with female pipe threads on the 1-inch end*
- (1) 2-inch-to-1-inch PVC reducing coupling, with female pipe threads on the 1-inch end†
- (2) 1-inch-diameter "close" cast-iron pipe nipples‡
- (1) 3-inch white PVC coupling, one side smooth, one side threaded
- (1) Can PVC primer
- (1) Can PVC cement
- Plumber's Teflon tape
- Duct tape
- (1) 1-inch-diameter U-shaped water sprinkler solenoid valve (Orbit Irrigation Model 57624 or equivalent, often available at hardware stores)
- (1) 3-inch-diameter white PVC threaded end cap
- (1) Air tank valve, ⅛-inch, NPT§
- (1) O-ring¶

---

* A 3-inch-to-1-inch reducing coupling and a 2-inch-to 1-inch reducing coupling is hard to find at hardware stores. You may substitute combinations of other available fittings, as long as you are ultimately able to go from a 3-inch-diameter PVC pipe to the 1-inch-diameter pipe nipple.

† These reducing fittings can also be hard to find at some hardware stores. You may substitute other available fittings, as long as you ultimately are able to go from a 2-inch-diameter PVC pipe to the 1-inch-diameter pipe nipple.

‡ A "close" pipe nipple is the shortest type of nipple and is threaded from each end with the threads meeting in the middle.

§ These are often available at large hardware stores. Alternatively, go to www.mcmaster.com and search for "⅛-inch tank valve."

¶ The O-ring seals the valve body to the end cap tightly against the flat boss on the 3-inch threaded end cap to prevent air leakage. Therefore, the diameter should be just large enough to slide onto the tank valve body. O-rings are available at most larger hardware stores.

The Electrical Assembly

- (3) 9-volt battery holders (e.g., Radio Shack part #2700326)
- (1) 2-inch-by-2-inch-by-5-inch project box (e.g., Radio Shack part #2701803)
- (3) 9-volt batteries
- (3) 9-volt battery wire connectors
- (1) Toggle switch with on-off label plate (e.g., Radio Shack part #275-602)
- (1) Momentary switch (e.g., Radio Shack part #275-609)
- Low-voltage insulated wire

General Materials

- Pressure gauge (like a bicycle tire gauge)
- Tie wraps
- T-shirt, towel, or other soft projectile
- Plastic bag
- (1) 3-foot-long, 1-inch-diameter wooden dowel (pushing stick)

MAKING THE T-SHIRT CANNON BARREL

6.1 Barrel Assembly

1. Cut the PVC barrel pieces. Measure and mark a cutting line 25 inches from one end of the 3-inch-diameter PVC pipe. Use the hacksaw to cut the pipe squarely.

2. Now, measure and mark a cutting line 32 inches from one end of the 2-inch-diameter PVC pipe.

3. Assemble a combination of reducing fittings that connects the 3-inch-diameter PVC pipe to the 1-inch-diameter pipe nipple. See **diagram 6.1.**

4. Similarly, assemble a combination of fittings for the 2-inch-diameter PVC pipe to the 1-inch-diameter pipe nipple. See **diagram 6.1.**

5. Solvent-weld the smooth end of the 3-inch threaded coupling to the other end of the 3-inch-diameter PVC pipe. See pages 11–12 for solvent-welding instructions.

6.2

6. Attach the pipe nipples at the ends of the barrels to the solenoid valve. Wrap the pipe threads with plumber's Teflon tape. Screw the barrel assemblies into the openings on the valve. (Some brands of solenoid valve may not be large enough to handle both barrels at the same time. There must be at least 3¼ inches of space from centerline to centerline in the solenoid valve to accommodate the barrels.) See **diagram 6.2**.

7. Wrap the 3-inch barrel with multiple layers of duct tape.

8. Drill a hole in the center of the threaded end cap just large enough to enable the air tank valve body to be screwed through the PVC of the end cap. You can thread the hole with a tap of

the same size as the threads on the valve body, or you can simply let the harder metal threads on the valve cut into the softer plastic. Once it's in place, slide the O-ring onto the valve body, and tighten the nut that comes with the valve. Tighten the nut down on the O-ring enough for a good seal, but do not overtighten.

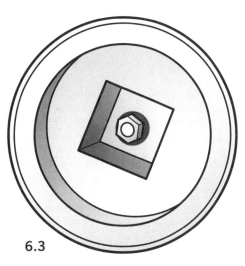

6.3

Detail of O-Ring,
viewed from inside cap

6.4

9. Screw the threaded end cap into the threaded end of the 3-inch-diameter coupling, using the Teflon tape to seal the connection tightly.

CONSTRUCTING THE ELECTRICAL ASSEMBLY

6.5 Electrical Assembly

1. Mount the battery holders in the project box. Place the batteries in the battery holders, and attach the 9-volt wire connectors to the batteries. Wire the batteries together in series to make a 27-volt power source. This will provide enough voltage to operate the 24-volt solenoid on the sprinkler valve.

2. In the top of the project box, drill a hole just large enough for the toggle switch (the safety switch) and another hole for the momentary-on push button switch (the trigger switch). Insert the switches through the holes, and secure them using the nuts supplied with the switches.

6.6 Electrical wiring

3. Wire the 27-volt battery supply, the toggle switch, and the momentary switch together in series. See **diagram 6.5**. Attach the solenoid leads to each end of the circuit.

4. Test the circuit by turning the toggle switch to "on" and depressing the push button. If wired correctly, the solenoid valve in the sprinkler valve will make a "thunk" sound when the momentary switch "trigger" is pushed.

ASSEMBLE THE T-SHIRT CANNON

1. When the circuit works correctly, solder all connections and close the project box. Attach the project box to the gun barrel using the tie wraps, as shown in **diagram 6.7**.

2. Allow all of the PVC connections to cure overnight.

3. Pressure-test the gun: using a bicycle pump or other pressurized air source, apply approximately 40 psi to the device through the valve at the end of the 3-inch barrel. Check all of the connections and attachments by brushing a soapy water solution on all of the joints. Bubbles indicate the presence of an air leak. Eliminate any leaks by tightening the pipe joints or reseating the O-ring. Properly constructed PVC joints should not leak. Leaks occurring at solvent-welded pipe joints are uncommon but may be difficult to fix. Try to remedy leaking joints by applying more solvent.

6.7

4. When the gun holds pressure and the solenoid valve operates when the trigger is pushed, the T-shirt Cannon is ready for use.

OPERATING THE T-SHIRT CANNON

6.8

1. With the safety toggle switch in the "off' position, load the cannon. Place the T-shirt in a plastic bag, or tightly wrap and fold the T-shirt so that it fits into the mouth of the barrel but is still slides fairly easily, and then secure it with tape. If the T-shirt unrolls during flight, it will not travel very far. Push the T-shirt down into the barrel with the pushing stick.

2. Using a bicycle pump or air compressor, pressurize the air chamber of the cannon to 40 psi. Check the pressure inside the tank with a pressure gauge. Using higher pressures will cause the T-shirt to travel farther but it's important to keep in mind that this is an "off-label" use of PVC pipe and this application has not been approved by the pipe manufacturers. Remember, you must build and operate this device at your own risk.

3. Move the safety toggle switch to the "on" or "fire" position. Carefully aim the cannon, and press the red trigger button.

PROBLEMS?

If the T-shirt Cannon doesn't fire, or if it fires weakly, try the following fixes:

- Make sure the air tank does not leak and the air pressure is high enough to shoot the T-shirt.
- If you suspect a leak, spray soapy water on the joints. You'll see bubbles wherever a joint leaks.

- If you cannot accumulate pressure in the air tank, check the position of the manual release switch on the solenoid valve to make sure it is closed. Make sure all pipe connections are tight and the O-ring seals securely.
- Make sure the solenoid valve is wired to the trigger button correctly. Remember, you should be able to hear the solenoid operate when the "fire" button is pushed.

THE ARCHITRONITO, LEONARDO DA VINCI'S STEAM CANNON

Leonardo da Vinci's steam cannon, the Architronito, was first described in detail in one of the many notebooks da Vinci kept for jotting down his ideas. Apparently, he was inspired by the Greek scientist Archimedes, who first devised a steam cannon to protect his home city of Syracuse from invaders. Archimedes's weapon used steam power to hurl a hard projectile with nearly as much velocity and momentum as a gunpowder-powered cannon.

The Architronito model described here is simple, consisting of a length of iron pipe, a few pipe fittings, a water injector, and a heat source. The really interesting thing here is the incredible amount of energy contained in the volume of steam produced from a very small measure of water. When water flashes to steam inside the steam cannon, enough energy is produced to easily propel a steel ball across the backyard. Imagine the power available from a full-sized steam cannon!

Leonardo da Vinci, although most well known today as a painter, actually made his living as a military engineer. During his lifetime (1452–1519), a number of city-states on the Italian peninsula, including Naples, Siena, Milan, and Florence, warred with one another. Because of the turbulent political environment, clever engineers were in high demand. Da Vinci designed on paper a number of highly original, novel weapons, but most were never used by combatants. In Leonardo's notebooks are many drawings and sketches, including designs for the first flintlock rifle, a 135-foot-wide rock-throwing ballista, and a rapid-firing repeater crossbow. Of these ideas, perhaps none is as interesting as the one for a steam-powered cannon.

Why was Leonardo interested in developing such a device? At the time, the exact composition of gunpowder was a closely guarded secret, and much of the powder was of poor quality. The availability of powder was also a problem. Gunpowder is a mix of sulfur, charcoal, and potassium nitrate (also called saltpeter). In gunpowder the sulfur and charcoal burn with oxygen—not from the air, but from the chemical bonds within the saltpeter.

In Renaissance Italy saltpeter was a rare and therefore expensive commodity. In the medieval period its only known sources were dung heaps. Only in the 19th century were alternative sources—guano (also dung, but from sea birds) from South America and salt lakes in Chile—were discovered. Today we make saltpeter using industrial chemicals such as ammonia and nitric acid, so no one has to go scrabbling around in sewage.

Because of the difficulty and cost involved in making gunpowder, Leonardo sought an alternative. His steam cannon design placed the cannon's breach (the closed end) in a pot of hot coals. An iron ball was placed in the muzzle of the cannon. Then a small amount of water was introduced, through a system of injectors and vents, into a small but extremely hot gap behind the ball. The water would flash to steam, expanding thousands of times in an instant. With a great hiss and whoosh, the steam would hurl the cannonball in the direction of the enemy.

Leonardo wrote about the quantities of water required, the size of the projectiles, and how far they could be fired. As it turns out, only a small quantity of water was required to hurl the projectile. Steam is a very energetic source of motive power for cannonballs. Based on the data in his notebook, it appears that da Vinci performed some experiments on the cannon. Most likely, though, the weapon was never put into battle, as there are no historical records describing its use.

Preceding Leonardo's steam cannon machinations was the device designed by the brilliant and eccentric geometer Archimedes. A catapult that used steam power, it was, in a manner of thinking, a cannon. Leonardo learned of it from a manuscript written by the Roman author and statesman Cicero. Italian scholar Francesco Petrarch (1304–74) found the manuscript in a church library. Petrarch collected Greek and Roman manuscripts that had been neglected for many centuries in various libraries.

The Cicero manuscript proved to be of great interest to Leonardo, who used the information to sketch the device he called the Architronito in honor of Archimedes. Apparently he produced only drawings of the steam gun.

In 1981 Ioannis Sakas, a Greek expert on the work of Archimedes, used this information to build an accurate replica consisting of a steam chamber, valve, and muzzle.

Sakas built a hot fire under the steam chamber. When the temperature inside reached 400° Celsius, Sakas introduced a very small amount of water, which he found produced enough rapidly expanding steam to throw a tennis-ball-sized stone 50 yards.

The steam cannon concept was not widely used in actual warfare until the British Royal Navy placed steam-powered grenade hurlers on board some of its ships during World War II. The Holman Projector was a 4-foot-6-inch-long pipe on a swivel mount connected to the ship's steam boiler through a pedal-operated valve. The general idea was that, once low-flying enemy aircraft began an attack, sailors would drop a live hand grenade down the pipe. They'd tug and manhandle two large iron handles to aim the swiveling pipe at the attacking Messerschmitt. At the instant the airplane swooped in, they'd stomp on the pedal. A tremendous blast of high pressure steam would gush through the valve body, propelling the hand grenade toward the enemy aircraft and, they hoped, knocking it out of the sky.

It was a clever idea—simple, easy to use, and cheap. But, most sailors who had occasion to fire the Holman Projector had a poor opinion of it. It was cumbersome to use and hard to aim. Many sailors thought the Holman was more dangerous to themselves than to the enemy. But there are reports of sailors shooting potatoes at other British ships as a form of recreation, possibly making the Holman the earliest-known embodiment of potato cannon technology.

The Architronito miniature steam cannon operates on the same principles as the inventions of Archimedes, da Vinci, and Holman. Basically, when water is heated in a closed reservoir, it changes to steam and exerts pressure on the container walls. When the valve to the reservoir is opened, the steam pushes on whatever object is in the barrel, sending it flying.

And fly they do. I used a ballistic chronograph to measure the speed of the projectile. Using 5 to 10 milliliters of water and about 30 seconds of heating time, I averaged a muzzle velocity of about 65 feet per second.

Safety

This project is not intended for children.

1. Heat is applied to the pipe using a propane torch in order to boil the water inside. The pipe becomes very hot. Do not touch it until it cools. Cooling can take several minutes.
2. Use no more than 5 to 10 milliliters of water.
3. The metal ball, along with the steam jet pushing on it, exits the barrel with substantial velocity. Clear the area in front of the cannon of people, animals, and breakable objects for 100 yards or use an acceptable shooting backdrop.
4. Never, ever look down the barrel of the cannon.
5. Wear safety glasses and heat-resistant gloves.
6. Handle the propane torch safely. Read and follow the propane torch manufacturer's instructions.
7. Your experience may differ from ours. Accidents happen; things can and do go wrong. Therefore, use at your own risk.

TOOLS

- Safety glasses
- (1) Metal-jawed vise
- Propane torch
- Heat-resistant gloves
- Pipe wrenches

MATERIALS

All pipe and pipe fittings are standard-wall iron or brass pipe with NPT (National Pipe Thread) threads. The pipe fittings normally available at most hardware stores are suitable.

- (1) 10-inch-long ⅜-inch diameter pipe nipple
- (1) 4-inch-long ⅜-inch diameter pipe nipple
- (1) ⅜-inch diameter pipe elbow
- (1) Ball valve with ⅜-inch NPT openings
- (1) Pipe reducer, ⅜-inch male NPT to ¼-inch female NPT
- (1) 2-inch-long ¼-inch diameter pipe nipple
- (1) ¼-inch pipe tee
- (1) Pressure gauge, ¼-inch NPT connection
- (1) Pressure relief valve, ¼-inch NPT connection
- Pipe sealing compound
- A quantity of ⁷⁄₁₆-inch steel balls. Steel balls can be ordered online from places such as McMaster-Carr (www.mcmaster.com) or Amazon.com

MAKING THE ARCHITRONITO

7.1

1. Connect the pipe fittings and valves together as shown in **diagram 7.1**. Use pipe sealing compound at each joint. It is

very important that all fittings are screwed together tightly with no leaks. Poorly connected joints will leak hot steam and cause poor performance.

2. Find a safe, suitable area for testing your steam cannon.

7.2

3. Securely clamp the cannon in a vise. Angle the barrel at least 15 degrees from the horizontal, but not more than 45 degrees.

4. Open the ball valve. Pour in 5 to 10 milliliters of water, making sure it reaches the reservoir section behind the valve.

5. Don safety glasses and heat-resistant gloves.

6. Close the ball valve. Insert a steel ball in the barrel. Review the safety instructions on aiming the cannon on page 104. Use the propane torch to apply heat to the reservoir portion of the cannon. As you heat, watch the pressure gauge while looking carefully for leaks around the joints. If leaks are present, wait for the device to cool before fixing the joint. Stop heating when the pressure gauge reads 20 psi. Do not exceed 20 psi.

6. If no leaks are present, fire the cannon by quickly opening the ball valve.

BOMBASTIC BLOWGUNS

There are myriad ways to shoot projectiles besides using nitrogen-laden chemical propellants. We've explored projectile shooters powered by springs (the Clothespin Snap Shooter), compressed air (the T-Shirt Cannon), and steam (the Architronito). The following projects use the most personal power source of all, air from human lungs.

The Marshmallow Shooter

The Marshmallow Shooter has probably been around since mini-marshmallows were invented. This is a favorite of schoolchildren everywhere. Simple to build and fun to use, this project will appeal to children of any age.

Safety

There aren't too many safety issues with this project. Wear safety glasses. Don't shoot people in the face with the marshmallows. Don't eat dirty marshmallows. That's about it.

TOOLS

- Safety glasses

MATERIALS

- (1) 10-inch-long, ½-inch-diameter PVC pipe
- (6) 4½-inch-long, ½-inch-diameter PVC pipes
- (2) ½-inch-diameter PVC end caps
- (2) ½-inch-diameter PVC tee joints
- (2) ½-inch-diameter PVC elbow joints
- (1) ½-inch adapter, one side threaded, one side smooth
- (1) ¾-inch-to-½-inch coupling, the ½-inch side threaded, the other smooth
- Bag of mini-marshmallows

BUILD THE MARSHMALLOW SHOOTER

8.1 Marshmallow Shooter Assembly

1. Fit all the parts together as shown in **diagram 8.1**. Note that no PVC cement is required or desired. Simply dry-fit the pieces together.

SHOOTING THE MARSHMALLOW SHOOTER

1. Remove the mouthpiece from the gun.
2. Load a mini-marshmallow (or two) into the barrel. Replace the mouthpiece.

3. Bring the gun up so that you can blow into the mouthpiece. The mouthpiece swivels so you can adjust it to suit your preference.

4. Blow a sharp burst of air into the mouthpiece. The bigger your huff, the farther the marshmallow will go.

8.2

THE PHYSICS OF BLOWGUNS

The blowgun converts human energy into projectile motion. Inside the barrel of a blowgun, an expanding push of breath pushes against the fins of a dart, causing it to accelerate down the tube.

A question that people sometimes ask is, How long should the blowgun be in order to obtain the greatest velocity? That's a good question, and one we'll attempt to answer. At what point does the dart reach its greatest speed?

Some people believe the dart is traveling fastest in the barrel just after the mouthpiece. They argue that the blast of air starts and stops suddenly, like a trumpet player playing a sixteenth note. That line of reasoning posits that after the initial push, the fins rub against the pipe walls, causing friction and slowing the dart until it clears the barrel.

In actuality, the amount of air huffed into the pipe is very large in comparison to the total volume of the tube. The volume of a 6-foot-long, ½-inch-diameter blowgun is

volume of blowgun = π (.25)^2 × 72 = 14 cubic inches

In anatomy studies scientists have found that the average volume of air in a medium blowgun huff is more than 60 cubic inches. For experienced blowgun hunters the huff is likely much greater. In fact, the huff from a typical hunter's lungs is so large in comparison to the tube volume that the dart accelerates continually down the entire length of the pipe, reaching maximum velocity as it leaves the muzzle.

So would a 10-foot blowgun be a better weapon? Would a 15-foot gun? The short answer is yes. The long answer is maybe. A hunter would impart greater velocity and accuracy to his dart with a longer blowgun. But he or she must figure out a way to conveniently carry it while hunting and must figure out a way to build such a tube out of a material strong and stiff enough to keep it from drooping and wobbling.

The Yagua Blowgun

Blowguns have been used for hunting small mammals, birds, lizards, and even fish for thousands of years. These devices are known throughout the world, although the cultures of the Amazon basin and the Malay archipelago are most strongly associated with them.

In the hot, steamy rain forests, hunters learned to use whatever nature provided in the way of tools, food, and weapons. And they found blowguns well suited to their needs. Perhaps because of the heavy vegetation and relative smallness of prey animals, the highly accurate but low-power weapon became a favorite method of hunting.

Blowguns go by a variety of names. In England, where they are used purely for sport, they are known as *blowpipes.* In South America the gun is called the *pucuña.* The jungle residents of Southeast Asia refer to them as *sumpitan.* Japanese ninja warriors called them *fukiya*, and Cherokee Indians gave them the name *tugawesti.* In all these places blowgun hunters took this activity very seriously, practicing until they could hit a small, fast-moving target at incredible distances.

Although we won't go as far as hunting prey in the jungle, the next project does have a wild, exciting flavor. Carefully note the cautions associated with the construction and use of the blowgun. Remember, a blowgun is not a toy. It is a hunting tool of considerable power, a demonstration of physics, and a display of cultural anthropology.

A blowgun is simple to build. Blowguns in South America and the Pacific are usually fitted with a mouthpiece, which can be a simple widening of the breach end, a separate funnel-shaped wooden cup, or a piece of hollow bone. The purpose of the mouthpiece is to concentrate the entire force of the blow behind the dart.

TOOLS

- Hand saw
- Utility knife or sandpaper
- Scissors

MATERIALS

- (1) 4-foot-long, ½-inch-diameter PVC pipe
- (1) ¾-inch to ½-inch smooth socket PVC reducing fitting
- (1) ¾-inch smooth to female pipe thread adapter
- (1) 1-inch straight to ¾-inch male pipe thread PVC reducing fitting
- (1) Can PVC primer
- (1) Can PVC cement
- (1) 5-inch-long wire cut from coat hanger, or a cotton swab with the cotton ball removed from one end*
- (1) 2½-by-2½-inch square of thin cardboard (such as manila folder cardboard)
- Transparent tape

---

* Paper swabs work adequately, but better performance is obtained using a longer, wooden swab. Wood-shafted cotton swabs are available at some drugstores as well as larger online drugstores. If creating a stun dart, you will also need 1 wire nut.

BUILD THE YAGUA BLOWGUN

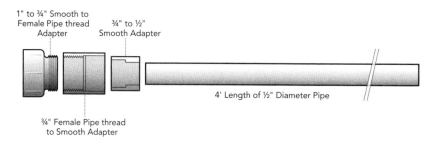

8.3 Yagua Blowgun Assembly

1. Using the handsaw, cut the PVC pipe to a 4-foot length.
2. Smooth both ends of the cut pipe with a utility knife or sandpaper, removing burrs.
3. Using PVC primer and cement, solvent-weld the ½-inch socket end of the reducing fitting to one end of the pipe. Solvent-weld the ¾-inch socket end to the smooth end of the ¾-inch smooth to female pipe thread adaptor. See page 11 for instructions on solvent-welding PVC pipe.

8.4

4. Screw the 1-inch to ¾-inch male pipe thread reducing fitting onto the ¾-inch female threaded fitting.

BUILD THE BLOWGUN DARTS

Stun darts are relatively heavy, have blunt ends, and don't pierce skin or animal hide upon impact.

Stun darts are easily made out of a wire nut and wooden cotton swab stick. A wire nut is a plastic electrical connector commonly sold at hardware stores. First choose a wire nut that fits the inside diameter of the barrel as closely as possible. Then insert a wooden cotton swab stick into the nut socket and twist it into place. The stick will add stability during flight.

Wire darts have thin, hard shafts and are used in small-game hunting.

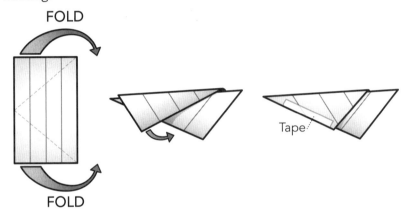

8.5

1. Roll the cardboard into an extended cone as shown in **diagram 8.5**. Wrap the cone with transparent tape to keep its shape.
2. Use the scissors to cut a hole in the tip of the cone that is the same diameter as the coat-hanger wire or cotton swab.

8.6

3. Insert the dart body into the hole. Tape the cone to the dart body securely, with the open end of the cone at the back of the dart.

8.7

4. Place the dart in the blowgun muzzle. The back of the dart will extend beyond the muzzle. Trace the muzzle outline on the cone using a marker. Remove the dart and cut the cone along the tracing as shown in **diagram 8.7**. Now the cone should fit the inside of the blowgun exactly. You may cut the cone slightly if necessary to ensure that the dart travels freely within the blowgun.

5. The quality of the dart is very important. The cone should be squarely attached to the wire or cotton swab and centered on it exactly. The cone must fit the inside of the barrel as closely as possible without sticking.

8.8

USING THE BLOWGUN

1. Look down the barrel to make certain it is clear and free from all obstructions.
2. Unscrew the mouthpiece and insert the dart. Replace the mouthpiece. **See diagram 8.8.**
3. Turn your head away from the mouthpiece and take a deep breath. Then, while holding your breath, bring the mouthpiece to your lips. Blow a quick hard burst of air into the blowgun. Never inhale through the blowgun.
4. Aiming takes practice. Most blowgun users leave both eyes open while they aim and shoot.
5. Blowguns can be dangerous. Always use responsibly!

ALEXANDER VON HUMBOLDT AND THE FLYING DEATH

The small diameter and relative lightness of blowgun darts mean that the projectile must be somehow augmented to be useful to a hunter. No animal larger than a mouse could be hunted effectively by blowgun unless the lethality of the dart is increased. The most obvious way of doing so is to use poison darts.

To be useful, the poison must have two important qualities. First, it must be fast-acting: no hunter wants to track a wounded animal all day, waiting for it to finally keel over. Second, the meat from the animal must remain safe to eat. Eighteenth-century Europeans knew that native hunters in South America used such poisons, but the composition of the toxin was not known until the turn of the 19th century, when Alexander von Humboldt solved the mystery.

While the great American explorers Meriwether Lewis and William Clark were mapping the American Northwest, another great explorer was doing the same thing about 4,000 miles to the south. A Prussian baron, assisted by a French doctor, had set out on a grand and ambitious mission to explore previously little-known areas of Central and South America.

Burning with a passion for increasing his personal store of knowledge, Baron Alexander von Humboldt, assisted by his friend Dr.

Aimé Bonpland, left La Coruña, Spain, on June 5, 1799, on a ship bound for the New World. They spent the next five years investigating, exploring, cataloging, and documenting the geography, geology, flora, and fauna of Cuba, Venezuela, New Granada (now called Colombia and Ecuador), Peru, and Mexico.

The influence of the Humboldt expedition's discoveries on the sciences of biology and geography is hard to overestimate. Humboldt and Bonpland collected a vast number of plants and animals. They took rock samples while they mapped and charted. They paddled up the Magdalena River in Colombia, then continued on foot across the rugged Andes to Lima, Peru. Humboldt and Bonpland attempted to summit what was at the time the highest Western-known mountain on Earth. (Everest was not yet known to non-Himalayans.) Humboldt climbed to within 1,400 feet of the summit of present-day Ecuador's highest peak, Chimborazo, but was stopped by an impassable vertical wall. Still, the expedition reached an elevation of nearly 22,000 feet. For the next 36 years Humboldt and Bonpland held the Western record for attaining the greatest height above sea level.

Humboldt and company were the first Europeans to see an electric eel, to map the ruins of ancient Incan cities, and to verify the existence of a small but important body of water called the Casiquiare Canal. The Casiquiare Canal is the only natural canal in the world that connects two major rivers, the Orinoco and the Negro Rivers.

The party wended its way 1,800 miles up the Orinoco River, charting it as they went. Perhaps equally important was Humboldt's work measuring the speed and direction of the ocean current off the coast of Peru, now called the Humboldt Current. A cold ocean current that runs along much of the western coast of South America, the Humboldt Current has tremendous effects on the ecology, weather, and economy of the entire Pacific coast.

No doubt about it, this trip was very difficult. At times the explorers had no food, and the remote, humid jungle conditions were almost unbearable. But the knowledge they gained was more than worth the hardship. Arguably, the best-known single episode during the trip occurred when Baron von Humboldt became the first European to witness South American Indians extracting and preparing a poisonous compound called curare. Up to then curare's composition was a secret known only to the inhabitants of Amazonia. For as long as the tribes could remember, they had smeared the tips of their blowgun darts with curare.

Known as the Flying Death, curare's action is swift and deadly. A simple prick from a curare dart is enough to cause death within minutes. So powerful is the toxin that it can kill a small animal in seconds, a monkey in 5 minutes, and a grown man in 10.

In his account of the experience, Humboldt wrote that the natives used the Flying Death for "war, the chase and gastric obstruction." Now, that information wasn't new, for references to the drug go as far back as the voyages of Sir Walter Raleigh, and other explorers had posted accounts of the legendary South American poison.

What was new was that Humboldt identified the plants from which curare was made and described what was done with them. The people of the area, the Ticunas, showed him the curare plant, which modern scientists refer to as *Strychnos toxifera*. He also found that the poison could be safely transported in bamboo tubes, that it could be tasted without danger, and when so used was an excellent aid for stomach trouble.

He watched and recorded as the Ticunas soaked the curare plant's roots, removed its bark with a knife or machete, and pressed out the juice with their hands. Mixed with some other compounds for handling, the juicy extract was carefully dried. While hunting, the natives would take small arrows or darts and dip the sharp tips in the curare.

Curare is a very powerful and dangerous poison. The hunters used a long, hollow blowgun to shoot a single, tiny, deadly dart at prey. Very often the target was a monkey. The potency of the Flying Death varied. If the monkey, after being shot with the poison dart, fell down after hopping from one tree to a single other, the Ticunas proclaimed it "one-tree" curare. If the monkey made it to a second tree before taking its fatal plunge, it was "two-tree." And if the curare wasn't quite up to par, it was "three-tree" poison, which entailed a lot of running around by the hunter to find and gather the paralyzed prey.

The curare Humboldt brought back was carefully analyzed. Since that time the drug has been used extensively in Western medicine, and it continues to be used in a variety of medical applications, such as muscle relaxants and anesthetics.

After returning home, Humboldt led a long and active life. He wrote extensively of his expeditions and discoveries. One of his books, *A Personal Narrative,* was important to young Charles Darwin and may have inspired him to take up his studies in biology.

When he was 60, Humboldt traveled to Siberia and Central Asia to study the weather. He died at age 90 and was buried in Tegel, Germany. Today Humboldt is widely remembered as a great explorer and geographer. Fittingly, several landmarks in the Americas bear his name, including the Humboldt Current, river, and mountain range. His largest monument is the Mare Humboldtianum, a lunar sea on the northeastern face of the moon.

XYLOEXPLOSIVE DEVICES

The ancient Greek word for wood is *xylon*. If you look through the dictionary, you'll see quite a few words in which the prefix "xyl-" has been added to other word roots to produce words related to wood. First of all, there's xylophone, which is a musical instrument made of wood strips. There's xylem which is the part of a tree in which the sap flows, and there's also xylene, which, like turpentine, is a type of paint thinner originally made from wood.

Perhaps you knew some or all of those words. But, here's a new one: xyloexplosive. The first part of the word, xylo, you now know means derived from wood. The second part, explosive, well, you already know what that means. So, logically, a xyloexplosive is a thing made of wood that explodes. Of course, wood can't explode like a bomb made from chemical explosives, but you can build a wood structure that is built from many small, tensioned wooden springs. If you build the xyloexplosive structure according to the directions that follow, the releasing of the energy in one of those springs causes another spring to pop, which in turn begets another and another and so on in a chain reaction.

Another name for a xyloexplosive is a stick bomb, and they are great fun to build. Despite their rather dangerous-sounding name, these constructions are safe for children of (almost) any age. Be warned: they take a good deal patience and a steady hand to build. But setting them off is both exciting and satisfying.

THE SCIENCE OF CHAIN REACTIONS

Xyloexplosives are basically one long mechanically meshed chain reaction. The term "chain reaction" was first used in 1918 by Max Bodenstein. Bodenstein was a German university scientist who was one of the founders of a branch of chemistry called chemical kinetics. He found that sometimes, when molecules react, they not only form what are called products of combustion, but also form some other molecules as well. When these other molecules are "unstable" they cause the reaction to continue propagating, faster and faster until all the original reactants are consumed. Such fast occurring chemical chain reactions are the reason explosives like dynamite and TNT explode. For these compounds, said Bodenstein, once a reaction starts, it "cascades," making the reaction proceed at faster and faster speeds until the reactants simply explode.

Chain reactions occur in all sorts of systems, not just in chemistry. There are biological chain reactions, mechanical chain reactions, and perhaps the most well-known type of all, nuclear chain reactions. Indeed, the concept of the nuclear chain reaction and the inventions that followed it are one of the major influences on world history since the middle of the 20th century. The man who first conceived of how a nuclear chain reaction could work was a young, relatively unknown European refugee living in London just before the outbreak of World War II.

LEO SZILARD AND THE ATOMIC CHAIN REACTION

Leo Szilard was born in Hungary in 1898. He was an exceptional scientist and worked closely with a number of top German theoretical physicists including Albert Einstein. When the Nazis rose to power, he left Germany for England. Although he wasn't rich, he had enough money to be able to take up residence in the Imperial Hotel near Russell Square in London. The Imperial was a pretty good hotel but what it was really well known for was its remarkable Turkish baths. Apparently Szilard did some of his best thinking in the bathtub, and the Imperial's baths were said to be very fine indeed.

Leo Szilard

Chemical chain reactions were well understood in 1933, but atomic chain reactions were not. In fact, on September 12, 1933, the *Times* of London published an interview with Ernest Rutherford, England's best-known physicist, in which he said that energy producing nuclear chain reactions were simply not feasible. The nuclei of atoms, said Lord Rutherford, could not be split, at least not in the foreseeable future. Atomic nuclei were made from positively electrically charged particles called protons. So if a scientist aimed one proton at a nucleus consisting of other protons in an attempt to split it, the electrical charges would simply repel each other and that would forever prevent the direct hit required to bust an atom apart.

Scientists of the time could see no way around the problem of similar charges repelling each other. So, most scientists agreed with Rutherford—the idea that atomic reactions could be used for making electric power or nuclear bombs was impossible, or in Rutherford's words, "moonshine."

On the rainy September morning that Szilard read the account of Rutherford's moonshine speech in the *Times*, it made him uncomfortable and he was actually even a bit annoyed. Szilard was a contrarian by nature, and moreover, he was the sort of person who became irritated when anybody, even a respected expert such as Lord Rutherford, made blanket pronouncements without plenty of rigorous scientific evidence. Piqued at Rutherford's scientific hubris, he decided to go out for a walk and think things over. Was the concept of atomic energy really just moonshine?

When he stopped for a red light on Southampton Street, he was hit with a flash of insight that changed the world. It occurred to him in an instant that what Rutherford said of charged atomic particles probably was true. But what, imagined Szilard, if the recently discovered chargeless atomic particles called neutrons were used?

At that street corner, Szilard first saw how electrically neutral atomic particles called neutrons would have no trouble getting by the electrical barriers inside atomic nuclei: "As the light changed to green and I crossed the street, it suddenly occurred to me that if we could find an element which is split by neutrons and which would emit two neutrons when it absorbs one neutron, such an element, if assembled in sufficiently large mass, could sustain a nuclear chain reaction."

This new way of looking at things fired Szilard's imagination like no other idea he had ever come across. Chain reactions became Szilard's obsession. He thought constantly about the way atomic chain reactions could occur and on March 12, 1934, a mere five months after Rutherford's moonshine speech, Szilard put his now fleshed out ideas on paper, filing a 15-page British patent application.

Shortly thereafter, Szilard began his collaboration with other top atomic physicists such as Enrico Fermi and in relatively short order, they were able to clearly understand the method in which chain reactions work at the atomic level. They found that inside atoms of uranium, the energy contained by one very, very fast moving neutron could split apart another uranium atom causing its neutrons to hit neighboring atoms. These "smashed atoms" in turn would emit neutrons that would smash still more atoms. Once the chain reaction began, it would continue, growing larger and larger until all the uranium was split and transmuted into other elements. But beyond transmutation (which means changing one element to another) atomic chain reactions release great scads of energy.

Such chain reactions are at the heart of the process called nuclear fission, the self-sustaining chain reaction in which fast moving neutrons create other fast moving neutrons that are the driving force inside nuclear reactors and atomic bombs.

Xyloexplosive Devices

What's true for atomic bombs is true for the mechanical chain reaction at the heart of xyloexplosives. Within our stick bomb, when the energy locked inside the wood fibers of a bent tongue depressor is set free, the energy released causes them to jump apart, or, in a manner of speaking, explode. And, in similar fashion to uranium neutrons busted loose in a fission reaction, there's enough energy in those exploding sticks to cause other tensioned sticks to pop loose. These in turn will cause other sticks to explode, and so on and so on, in a cascading chain of events that you can call a xyloexplosive chain reaction.

Popsicle sticks are thicker but narrower while tongue depressors are wider and thinner. Popsicle sticks explode with more energy, and leap higher when set free, making the reaction go very quickly. Xyloexplosives made from tongue depressors are easier to use and less likely to go off accidentally, but the chain reactions proceed with somewhat less speed and energy.

Safety

Unlike the energy inside uranium atoms (and some of the other projects in this book), the energy produced by a xyloexplosive device is tiny and occurs over time, so making one is a safe project for kids of all ages. But, I do recommend that all xyloexplosive makers wear safety glasses, especially when using thicker Popsicle sticks, as you never know where a flying stick could go.

TOOLS

- Safety glasses

MATERIALS

- (1) Large box of Popsicle sticks, craft sticks, or tongue depressors

SQUARE WEAVE XYLOEXPLOSIVE DEVICE

The simplest stick bomb uses a repeating pattern of sticks placed 90 degrees to one another in what is generally referred to by stick bomb builders as a square weave. To build the square weave, follow the instructions here. Note that the sticks run in a continuous over/under pattern as shown in **diagram 9.1**. For example, when the outside east-west running sticks are on top, you'll need to weave a north-south stick under them and over the center east-west stick.

1. Make a "T" with the two sticks, with the east-west stick on top of the north-south stick. **(9.2)**

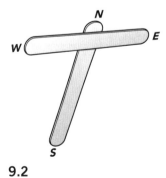

9.2

9.1 Square Weave Xyloexplosive Device

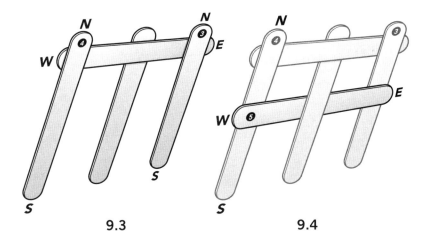

9.3   9.4

2. Place the ends of two "outboard" NS sticks on the ends of the EW stick, with the NS sticks on top. **(9.3)**

3. Weave an EW stick under the middle NS stick from step 1 and over the outboard NS sticks added in step 2. **(9.4)**

9.5   9.6

4. Place the end of another NS stick on top of the middle NS stick about ¼ from the end. **(9.5)**

5. Place two more NS sticks in parallel under the two outboard NS sticks. **(9.6)**

9.7　　　　　　**9.8**

6. Weave an EW stick over the middle NS stick added in step 4 and under the outboard NS sticks added in step 5. **(9.7)**

7. Place the end of another NS stick under the middle NS stick added in step 4. **(9.8)**

9.9

8. Place two more NS sticks in parallel on top of the two outboard NS sticks added in step 5. **(9.9)**

9. Continue weave by repeating steps 3 through 8 until you've attained the length desired or run out of sticks.

10. Insert a locking stick by weaving it over the final center NS stick and under the outer side NS stick. **(9.10)**

FIRING THE SQUARE WEAVE

To start the chain reaction, quickly remove the anchor piece that closes off the end of the weave (stick 14 in **9.10**) by flicking it with your finger. When the anchor is flicked away, the reaction commences and the sticks fly apart. This continues down the weave until the tensioned under-over pattern is completely eliminated.

9.10

COBRA WEAVE XYLOEXPLOSIVE DEVICE

A fascinating variant of the square woven stick bomb makes use of a crisscross pattern sometimes called a cobra weave.

1. Place Stick A underneath and at a 60 degree angle to Stick B. **(9.11)**

9.11

9.12   **9.13**

2. Place Stick C over the right end of stick B, parallel to stick A. **(9.12)**
3. Weave Stick D under Stick A and over Sticks B and C. **(9.13)**
4. Place the end of Stick E over the end of Stick A and under Stick C, keeping E parallel to D. **(9.14)**

9.14   **9.15**

5. Place the end of Stick F on top of Stick D, but under the middle of Stick E. **(9.15)**

6. Place the end of Stick G on top of Stick C, but under the middle of Stick F. **(9.16)**

9.16

9.17

7. Place the end of Stick H on top of Stick E, but under the middle of Stick G. **(9.17)**

8. Repeat steps 5, 6, and 7 with additional sticks until you've attained the desired length or run out of sticks.

9. Lock the weave by placing a locking stick on top on top of the two outboard sticks but underneath the middle stick. **(9.18)**

FIRING THE COBRA WEAVE

Start the chain reaction by flicking away the locking stick. If created skillfully, the cobra weave stick bomb takes on the aspect of a hooded cobra getting ready to strike when it goes off.

9.18

THE WHOOSHING
SKY PUPPET

Y ou certainly have seen those wacky, waving, inflatable tube guys, also known sky puppets or air dancers. The long-bodied, anthropomorphic plastic beings whoosh around their base in a delightful, animated fashion, moving in and out like a punch drunk boxer.

It may seem as if they've been whooshing around forever, but the sky puppet first blew into public consciousness at the 1996 Olympics in Atlanta. Since then, they've epitomized low-brow advertising and become a staple at used-car lots, check-cashing stores, and fireworks stands.

Behind the goofiness, there's a lot of science going on inside that tube. If the blower connected to the air puppet is producing too much or too little air flow for its length and diameter, the puppet is either an uninteresting flaccid tube of plastic or a spasmodic column of chaotic air.

But at the right airflow, just above the level when the molecules of air racing through the tube change from laminar to turbulent flow, the sky puppet becomes perfect eye candy, straddling the border between hilarious and mesmerizing.

AIR DANCERS, SKY PUPPETS, AND WACKY WAVING TUBE GUYS

To understand the physics of our creation, we must look to the late 19th century scientists who were just then becoming able to understand the intricacies of fluid mechanics. While it was relatively straightforward to apply Isaac Newton's laws of motion to solid objects like cannon balls and rockets, fluids like air and water were a different kettle of fish. Daniel Bernoulli and Leonard Euler were among the first to study how forces and energy caused fluids to move. They found that understanding the nature of fluids required a great deal of mathematics.

In 1883, British scientist Osborne Reynolds chugged through the complicated math and developed a ratio that predicted fluid flow patterns through pipes and ducts based on the ratio of fluid speed to viscosity. As long as the moving fluid's Reynold's number, as this ratio is known, is below 4,000, the flow is orderly. But over 4,000, the flow inside a pipe becomes turbulent. And that's just the thing desired in a wacky, waving, inflatable tube man.

Whooshing Sky Puppet

Building a five-foot-tall puppet only takes an afternoon and costs about $25 or possibly less, depending on what you already have lying around the house.

TOOLS

- Hammer
- Scissors
- Hot-glue gun
- Safety glasses

MATERIALS

- (2) Wood pieces, 12 inches x 12 inches x ¾ inch
- 1½ inch-long nails
- (1) 30 to 35 gallon plastic trash bag
- (1) ¾-inch roll of clear adhesive tape
- (1) 3-inch to 1½-inch PVC reducing fitting
- (1) 130 cfm or larger 12- to 18-volt DC electric blower with a 3-inch opening (Internet search for "bilge blower")
- (2) #10 round head wood screws, ¾-inch long
- (1) 12- to 18-volt battery, DC wall transformer, or power supply (You can wire two 9 volt batteries in series to make an 18 volt power supply.)
- (2) Alligator test leads

BUILD THE WHOOSHING SKY PUPPET

1. Build a stand for the blower by nailing the end of one board perpendicularly to the center of the other.

2. Cut the garbage bag into 2 pieces, each 3¼ inches wide by 2½ feet long. Form each piece into a tube using the ¾-inch tape to seal. Connect the pieces together using the ¾-inch tape. Leave both ends of the tube open.

10.1

3. Connect the wide end of the PVC fitting to the blower outlet using hot glue. With the outlet pointing straight up, affix the blower to the vertical board on the stand using the wood screws.

10.2

4. Connect one end of the garbage bag tube to the PVC reducing fitting.

5. Don safety glasses. Use alligator clips to connect the positive and negative battery terminals to the appropriate wires on the blower.

6. Enjoy the show!

10.3

PROBLEMS?

Adjust the waviness level by making the tube narrower, reducing or increasing its length, or decreasing or increasing the airflow by changing the voltage.

10.4

LITTLE LUDGAR, THE GRAVITY POWERED HURLING MACHINE

Around the turn of the 14th century the English king Edward I was leading his soldiers in a series of battles against Scotland. His zeal for battle would earn him his nickname "Edward, Scotto-Rum-Malleus," which means "Edward, Hammer of the Scots." Edward was a warrior king, an able field marshal, and a man who knew the value of using the latest technology to his advantage.

When he cornered a Scottish army inside the walls of Stirling Castle in southern Scotland, Edward ordered his engineers to get busy. He told them to build 13 of the most powerful gravity powered siege engines ever made. One by one in the spring of 1304, a phalanx of huge trebuchets were erected outside the walls of Stirling, under the dismayed and anxious gaze of the Scots garrisoned inside.

English soldiers cut down entire swaths of oak and beech forests to procure enough wood for the job. Then, waggoneers and teamsters brought cartload after cartload of timbers to the plains outside the castle walls. And there, the assembled craftsmen erected the machines.

These giant stone-throwing machines were given rather gentle sounding names that belied their power and destructive purpose—the Parson, the Forester, Belfry, Tout le Monde, the Vicar, and the Verney were just a few.

The biggest and most awesome machine was erected last, and it alone had a fearsome name. At the time, English kings spoke French, so Edward referred to the trebuchet as "Loup de Guerre"—the War Wolf. His army soon shortened it to a much more Saxon sounding "Ludgar." By any name it was a fearsome hurling engine and had the horsepower to fling projectiles that could destroy castle walls 12 feet thick. The stones it cast weighed as much as a small tractor.

By Easter 1304, many of the engines were ready and the artillery barrage of the castle began. By June all of the trebuchets, except for the gigantic War Wolf, were operating.

The siege lasted from April 22 to July 20, 1304. This is a surprisingly long time for the castle garrison to hold out considering the size and resources of the besieging English forces. The smaller catapults worked day and night battering the walls and hurling projectiles over the castle walls, but the Scots remained defiant.

Finally, in July, Ludgar was ready for action. The English army eagerly looked forward to seeing what this massive artillery piece could do. The excitement was so great that Edward's wife, Margaret, watched the final attack from a specially made tower with a viewing balcony.

Then something very disappointing happened—the Scots had had enough. Like the other trebuchets, Ludgar was constructed plainly and purposefully in their view. And this machine by all accounts was enormous. The Scots could see the gigantic beams and baskets being winched into place. They could see the stone missiles being arrayed for use against them as ammunition. They held council and talked it over, desperately seeking any possible way out, any way of escape. Finally they came to the obvious conclusion—they couldn't outlast the English, not with Ludgar and his mechanical friends pitching rocks at them all day and all night, and they couldn't get past the surrounding English army. So, they raised the white flag and sent a messenger with an offer of surrender to Edward.

Not so fast, replied Edward. Edward had spent lots of time and lots of effort on Ludgar, and he wanted to see something for his money. Edward ordered the unfortunate Scots back inside Stirling Castle and the siege was taken up again. With its first toss, Ludgar broke down an entire castle wall, crumbling it like a broken clay pot. The next tosses fell upon the buildings inside the castle ward and they collapsed into heaps of wreckage and debris. Finally, with the castle reduced to rubble, Edward accepted the Scots' surrender. At the end of the siege, only 30 men crawled out of the debris and Edward, say the chroniclers, spared their lives.

Little Ludgar

A trebuchet is a siege engine that uses a heavy mass, called a counterweight, to rotate a lever that in turn, flings a projectile high into the air and toward a target. A well-made trebuchet is accurate and powerful. The small gravity powered catapult described here was inspired by King Edward's original Ludgar. While much smaller and likely unable to knock down a bird house, much less a castle wall, it is a fascinating hurling machine with a bit of a modern look to it.

Little Ludgar uses a 10-pound counterweight made from barbell weights. This design uses plastic pipe mounted to a stable wooden base and is triggered by pulling on a wood prop that supports the counterweight. The downward motion of the falling weights causes a lever to flip from one side of the trebuchet to the other, and as it does so, a sling attached to a golf or tennis ball is flung high and far, toward an imaginary castle wall across your yard.

Safety

A counterweight-powered trebuchet is a fairly straightforward machine. Little Ludgar is constructed from PVC pipe affixed

137

onto a wooden base. PVC isn't a particularly strong material, but if you don't go overboard with the amount of weight, it gets the job done. But if you overload it (or sometimes, just experience bad luck) the structure could break, so make sure your hands and feet are never underneath the weights.

The only tricky part of this project is sanding down the interior of the two tee fittings so they slide easily down the guide pipes. Take your time and keep sanding until the tees slide as smoothly as possible along the guides. When sanding PVC parts, wear a dust mask so you don't inhale PVC particles.

Always wear safety glasses in case a pipe breaks and keep all parts of your body away from the falling weights and fast moving throwing arm.

TOOLS

- Rotary tool, such as a Dremel, with a sanding drum
- Dust mask
- Safety glasses

MATERIALS

- (2) 1¼-inch PVC cross fittings
- (1) 1¼-inch PVC tee fitting
- (2) 1¼-inch PVC flange fittings
- (2) 5-pound weight plates
- (2) 1¼-inch diameter PVC pipe, 26 inches long (uprights)
- (1) 1¼-inch diameter PVC pipe, 11 inches long (crossbar)
- (3) 1¼-inch to ¾-inch pipe bushings
- (3) ¾-inch PVC pipe caps
- (1) ¾-inch diameter PVC pipe, 20 inches long
- (1) Can PVC primer
- (1) Can PVC cement

- (2) 3-foot pieces of ⅛-inch diameter cord or twine
- (6) 1⅝-inch diameter rubber O-rings
- (2) 1¼-inch PVC pipe caps
- (2) ¾-inch diameter pipe, cut to a size to hold weight tightly against pipe bushing
- (1) 18 inches by 18 inches by ¾-inch plywood board
- (2) 2-inch by 4-inch by 8-inch wood boards (strike board support)
- (1) 2-inch by 4-inch by 16-inch wood board (strike board)
- (8) deck screws, 2½ inches long
- ½-inch diameter dowel, 18 inches long (prop)
- 8 feet of medium weight nylon twine
- (4) 1¼-inch by 1-inch long lag screws
- Small balls for ammunition

BUILDING LITTLE LUDGAR

1. First you need to modify the 1¼-inch PVC crosses so they can slide easily down the 1¼-inch PVC vertical pipes. In addition, you'll need to modify the 1¼-inch PVC tee fitting so it rotates freely on the falling cross arm. To do this, use a rotary tool such as a Dremel with a sanding drum to remove the ridge shown in **diagram 11.1** and enough PVC from the interior of the crosses and tee so the pipes slide through them easily. Sanding down the PVC isn't difficult, but it is a messy job, so wear a dust mask and safety glasses while you work.

Ridge

11.1

Nylon Twine

Modified
1¼" PVC Cross

Nylon Twine

1¼" PVC Pipe
11"

¾" PVC Pipe
cut to fit
weight snug

Modified
1¼" PVC Tee

1¼" to ¾"
Pipe Bushing

5 lb. Weight Plate

¾" PVC Pipe
cut to fit weight snug

¾" PVC Cap

¾" PVC Cap

20"

¾" PVC Pipe

11.3 Little Ludgar Crossbar
Assembly

¾" PVC Cap

2. Attach the pipe flanges to the plywood base securely using the ¼-inch lag screws. Then, dry fit the 26 inch upright pipes into the flanges as shown in **diagram 11.2**.

11.2

3. Dry fit the crossbar as shown in **diagram 11.3**. Then, slide the crosses onto the 26-inch uprights and adjust the fit until the sliding crossbar piece freely slides down the pipes. It's very likely that you'll need to make adjustments to the alignment of the PVC fittings.

4. When all the pieces are perfectly aligned and slide easily, permanently assemble your trebuchet by solvent-welding the parts together as shown in **diagram 11.3** using PVC primer and cement. See page 11 for instructions on solvent-welding PVC pipe. Do not solvent-weld the tee fitting!

5. When the weight falls during operation, great stresses are placed upon the sides of the outermost plastic crosses, the ¾-inch PVC pipes, and the pipe caps. I recommend using nylon twine wound in a crisscross manner to reinforce the crosses as shown in **diagram 11.3**.

Modified
1¼" PVC Cross

¾" PVC Pipe
cut to fit
weight snug

Nylon Twine
Reinforcement

¾" PVC Cap

1¼" PVC Pipe

5 lb. Weight Plate

1¼" to ¾"
Pipe Bushing

1¼" PVC Pipe

26"

Flange

11.4

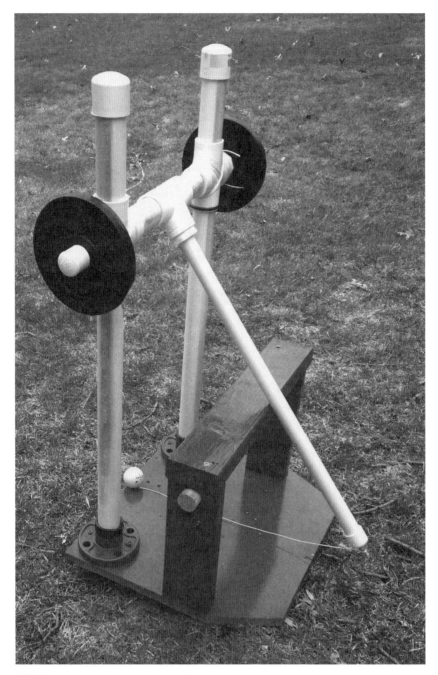

11.5

5. Place three O-rings around the bottom of each 26-inch pipe
   to absorb energy from the falling weights. The O-rings act as a
   cushion to the falling cross piece and prevent it from breaking.

5. Build the wooden strike bar from the 8-inch and 16-inch long two-by-fours and attach it to the plywood base using the 2½-inch long deck screws. The strike bar is merely a wooden bar that flips the throwing arm over when it falls. You can experiment with the positioning of the strike bar for best performance.

Expert tip: To get the most distance from your trebuchet, use a sling to connect your projectile to the throwing arm. A sling increases the length of the lever arm and gives you more throwing power. While medieval catapults used a sling attached to the throwing arm, it can be difficult to get them to work properly, so it's easier to simply permanently attach the sling to the projectile. To do this, securely tape a 3/16-inch wood dowel to the throwing arm so it extends over the top of the throwing arm by half an inch. Then tie a loop in the sling which you then hook over the pin atop the throwing arm, as shown in **diagram 11.5**. You can try golf balls, small water balloons, or other items of your own choosing for ammunition.

FIRING THE LITTLE LUDGAR

1. Put on safety glasses prior to operating the machine. Stay well clear of the rotating cross arm. Keep hands and feet well away from underneath the sliding cross arm. Do not stand in front or in back of the machine when the weight is elevated.
2. To unleash the power of your mighty hurling machine, insert the uprights into the openings of the crosses on the weighted crossbar. Place end caps on the uprights but do not use cement.
3. Raise the crossbar as far as possible and position the wooden prop to hold it up. Position the rotating arm on crossbar so that the strike bar will flip the arm and hurl the projectile. Tie the 8 foot rope to the prop.
4. To fire, stand well away from the hurler and tug on the rope to pull the prop bar away.

CONCLUSION

There are two people whose lives and works really encapsulate the culture of those who experiment with the projects and histories described here. Neither is very well known outside their small group of admirers, but they're both definitely worth knowing.

The first is Saint Barbara. Barbara is the patron saint of artillerymen, fireworks makers, bomb-disposal technicians, miners, and anybody else whose actions put them in contact with things that go boom.

Most of the stories surrounding the life of Saint Barbara are myths that have arisen out of earlier legends and folklore. According to one of many legends involving Saint Babs (as artillerymen who look to her for protection often call her), she was a young and pious woman who lived in the area of Nicomedia in fifth-century Asia Minor. In the summer of 430 an army of Vandals marching eastward from their Germanic homelands arrived at the walls of the city. The Nicomedians took one look at the size of the approaching army and quickly barricaded themselves behind the high, thick town walls. A long siege began.

The leader of the city, a man named Alypius, retrieved his daughter Barbara from the convent where she lived, to assist in the defense of the city. Alypius and Barbara took large jars and filled them with a mysterious substance known only to them. They ordered that the jars' contents be poured into the trenches surrounding the city, which then ignited into a towering inferno, an impenetrable wall of flame that repelled the attackers.

After a pause to regroup and reconsider, the Vandals continued their siege. Over the next year Barbara and her charges repeatedly repelled the attackers by using a variety of incendiary weaponry, including, so the legend says, burning globes of fire hurled from catapults.

In the end, though, the numerical superiority of the besiegers overcame the city's defenses. The besieging army stormed into the convent where Barbara had sought refuge. But the warrior-saint had a final trick up her sleeve. She had amassed a huge store of the mysterious explosive in a subterranean passage beneath the convent.

On Barbara's order, a gigantic explosion was set off, and the fireball consumed both the Vandal horde and Barbara; the Vandals were trapped beneath the debris of the exploded convent, and Barbara shot heavenward presumably to glory and sainthood.

Besides having their own patron saint, those who work with things that go boom have a patron philosopher as well. Georg Lichtenberg was a distinguished 18-century German scientist and scholar. Lichtenberg, while somewhat well known today, was an intellectual luminary of his time. He was a fellow of Britain's Royal Society, an esteemed professor at Göttingen University, and a friend and confidant of such leading lights as Immanuel Kant, Johann Wolfgang von Goethe, and Alessandro Volta. And he loved fast-moving, noisy things.

The polymathic Lichtenberg was an experimental physicist, astronomer, mathematician, and diarist. He is given credit for being among the first university instructors to incorporate experimentation into his classes, and he is the discoverer of xerography, from which modern copying machines descended.

But arguably, he is best remembered for his quotable aphorisms, all of which are short, clever, and razor sharp.

One of Lichtenberg's best-known quotes is a favorite of mine and makes for a fitting conclusion to this book. So as you attempt to build the projects described in the previous pages, take a moment to remember what Lichtenberg wrote regarding loud, kinetic, and energetic experiments:

> A physical experiment which makes a bang is always worth more than a quiet one. Therefore a man cannot strongly enough ask of Heaven: if it wants to let him discover something, may it be something that makes a bang. It will resound into eternity.

SOURCES OF MATERIALS

Most of the materials and tools used for the projects in this book are available at larger hardware stores and lumberyards such as Lowe's, Menards, or The Home Depot. If you cannot find a local parts supplier, several large industrial supply houses may be able to provide the parts you need. Supply companies such as McMaster-Carr Supply Company (www.mcmaster.com) have large inventories and well-organized websites.

A few projects require some basic electrical supplies such as switches, batteries, and wiring, which are available at stores such as Radio Shack or online at places such as sparkfun.com, amazon.com, or newark.com

Stun guns are widely available from suppliers on the Internet. A search for the word *stun gun* will return a large number of vendors. Amazon and Cabela's (www.cabelas.com) often list them for less than $30.

Transparent PVC, used mainly in the food-processing industry, is available from industrial plastic suppliers. Check the business listings in your telephone directory or search the Internet. Transparent PVC pipe is expensive when purchased at full price, but you may be able to purchase remnants and scrap at greatly reduced prices from a supplier or food-processing company.

ACKNOWLEDGMENTS

Thanks to the many readers of my past books who took the time to write me about their experiences and to share ideas. Contact me with comments or suggestions by visiting www.williamgurstelle .com and following the instructions you will find there.

My sincere gratitude goes to those who provided inspiration, project suggestions, and candid advice. Steve Cox, Jim Keifenheim, Jerry Pohlen, Caz Sienkiewicz, and my many collaborators at *Make Magazine* and *Popular Science* are first among them.

Thanks to Cynthia Sherry and Michelle Williams at the Chicago Review Press for their help, advice, and contributions. As always, my capable agent, Jane Dystel, provided great advice and service.

INDEX